Solid Air

Invisible Killer:
Saving Billions of Birds from Windows

with foreword by
Dr. David M. Bird

Daniel Klem, Jr.

Photographs of window-killed birds from Fatal Light Awareness Program (FLAP).

Copyright © 2021 Daniel Klem, Jr.

Cataloguing data available from Library and Archives Canada
978-0-88839-646-4 [paperback]
978-0-88839-640-2 [hardcover]
978-0-88839-665-5 [epub]

All rights reserved. No part of this publication may be reproduced, stored in a retrieval system or transmitted, in any form or by any means, electronic, mechanical, audio, photocopying, recording, or otherwise (except for copying permitted by Sections 107 and 108 of the U.S. Copyright Law and except for book reviews for the public press), without the prior written permission of Hancock House Publishers. Permissions and licensing contribute to the book industry by helping to support writers and publishers through the purchase of authorized editions and excerpts. Please visit www.accesscopyright.ca.

Illustrations and photographs are copyrighted by the artist or the Publisher.

Printed in the USA

Cover: Front cover photograph of Blackburnian Warbler (*Setophaga fusca*) by Kate Ekanem, William Keller, and Johnathan Schlegel.
Cover photograph was inspired by the art of Miranda Brandon.
Front and back cover designs by Robyn L. Klem

Production & Design: L. Raingam, J. Rade & M. Lamont
Editor: D. Martens

We acknowledge the financial support of the Government of Canada through the Canada Book Fund and the Canada Council for the Arts, and of the Province of British Columbia through the British Columbia Arts Council and the Book Publishing Tax Credit.

Hancock House gratefully acknowledges the Halkomelem Speaking Peoples whose unceded,

Published simultaneously in Canada and the United States by
HANCOCK HOUSE PUBLISHERS LTD.
19313 Zero Avenue, Surrey, B.C. Canada V3Z 9R9
#104-4550 Birch Bay-Lynden Rd, Blaine, WA, U.S.A. 98230-9436
(800) 938-1114   Fax (800) 983-2262
www.hancockhouse.com   info@hancockhouse.com

Solid Air

Invisible Killer:
Saving Billions of Birds from Windows

Daniel Klem, Jr.

For my parents Daniel and Lillian Mary (Perlis), parents-in-law Angelo Michael and Patricia A. (Deitrick) Mucci, my wife Renee A. (Mucci) Klem, and our children Heather Anne, Robyn, and Danny.

Thankful for your continuous and unlimited support.

"A thorough, important, and ultimately hopeful review of everything that is known about bird window collisions, by the world's leading expert in the field. This book is filled with useful information that should help all of us tackle this insidious problem."

— DAVID ALLEN SIBLEY, ORNITHOLOGIST, ILLUSTRATOR AND AUTHOR OF THE SIBLEY GUIDES TO BIRDS

It is not exaggeration or hyperbole to say that there is no one on the planet who has devoted as much time and effort studying the perilous relationship between birds and glass as Dr. Daniel Klem, Jr has and, consequently, few people can write as authoritatively and holistically about that relationship. *Solid Air* is an invaluable addition to the bird-glass collision literature and everyone—whether they are unfamiliar with the glass collision issue or already have an extensive background and understanding—will benefit from this immensely readable and absorbing book."

— JOHN ROWDEN, PHD SENIOR DIRECTOR, BIRD-FRIENDLY COMMUNITIES, NATIONAL AUDUBON SOCIETY

"Having worked with Dr. Daniel Klem, Jr for many years to build awareness of bird collisions in North America, we are delighted to see his unparalleled expertise come to literary fruition. Dr. Klem has brought his signature charm to this book, along with a wealth of information about bird strikes, their cause, and what can be done to prevent them. At once technically thorough and down to earth, Solid Air is a must-read for bird conservationists and sustainably-minded building professionals."

— CHARLES ALEXANDER VICE-PRESIDENT OF SALES AND MARKETING, WALKER GLASS

"The book is easy to read, accessible, and chronological. The reader is captivated by the unfolding story and the frustrations and mysteries explored by the author. *Solid Air* will appeal to the bird-watcher, scientist, conservationist, architect, and anyone who has ever heard that awful thud of a bird hitting a window. *Solid Air* reminds us this is a threat we can address and provides the tools to do so."

— LAURIE J. GOODRICH, PHD SARKIS ACOPIAN DIRECTOR OF CONSERVATION SCIENCE, HAWK MOUNTAIN SANCTUARY ASSOCIATION

"For architects, *Solid Air* is an essential contribution to the generalist knowledge that is critical to the creativity and competency of our profession."

— JULES DINGLE, PRINCIPAL, DIGSAU ARCHITECTURAL FIRM

"This is the book we have been waiting for. Daniel Klem has been the leading researcher on bird-window collisions throughout his long career, and thanks to him there is now widespread recognition of the magnitude and severity of the problem, and of measures that can reduce it."

— PROF. DAVID FRASER, ANIMAL WELFARE
PROGRAM, UNIV. OF BRITISH COLUMBIA

"If people were to follow Klem's guidance, the result would be hundreds of millions fewer bird collisions per year which would help stem the decline of our most vulnerable birds. For anyone interested in birds and bird conservation, as well as a rare insight into how scientists think and are motivated, this is a must-read."

— PAUL KERLINGER, PHD CAPE MAY, NJ AND
AUTHOR OF HOW BIRDS MIGRATE

"*Solid Air* is a must-read for any serious student of ornithology or anyone who cares deeply about bird populations. Dan Klem Jr.'s writing is thoroughly engaging as he takes the reader on three equal, interconnected journeys any one of which would be worth reading for their own sake. Purchase the book, read it, put knowledge into action and then share it with someone else who lives in a house with glass windows."

— PROF. JAMESON F. CHACE, DEPT. BIOL. & BIOMED.
SCIENCES, SALVE REGINA UNIVERSITY

"*Solid Air* makes the case: the discovery, the approaches to the problem, the tragic losses, the struggle for popular and the scientific concern, and the solutions available. Klem ultimately leaves us with the necessary tools to conquer this bird-and-glass issue. We only need to grasp the lessons and pursue the necessary changes."

— PAUL J. BAICICH, CO-EDITOR BIRDING COMMUNITY E-BULLETIN
AND REGULAR COLUMNIST FOR BIRD WATCHER'S DIGEST

"His writing is humble, personal, compelling and provides a stirring call to action for every one of us. A must-read for avian conservationists, architects, designers, glass manufacturers, community activists, and bird enthusiasts alike."

— MARY COOLIDGE, PORTLAND AUDUBON

"The concepts and guidelines outlined in this book must be part of every architect, designer, and planner's vocabulary and is critical to protecting our local and global ecosystems through responsible design and construction."

— KATE SCURLOCK, GWWO ARCHITECTS

Table of Contents

Foreword 12

Preface 16

Prologue 18

Chapter 1: Humans Make Glass, Birds Hit Glass 23

Chapter 2: Why Birds Hit Windows 28

Chapter 3: How Big a Problem 34

Chapter 4: Scientific Evidence – Making the Case 42

Chapter 5: What Happens When Birds Strike Windows –
Cause of Death,
Injuries, Recuperation 68

Chapter 6: Valuing the Dead 75

Chapter 7: Why do We Care – the Services Birds Provide 77

Chapter 8: Getting the Word Out – Transforming Education into Action ... 81

Chapter 9: Solutions – Solving the Problem 107

Chapter 10: Legal Protection of the Vulnerable and Defenseless 134

Chapter 11: What Citizens Can do as Homeowners and Advocates 144

Chapter 12: Seeing the Invisible Threat – Overall Impact 152

Resources 155

Bird-Window Collision Prevention and Cost 157

Bird-Window Collision Species by Geography 160

Bird Species Known to Strike Windows 162

Bibliography 163

Acknowledgments 199

Index 209

About the Author 221

The Glass in the Field

A SHORT TIME AGO SOME builders, working on a studio in Connecticut, left a huge square of plate glass standing upright in a field one day. A goldfinch flying swiftly across the field struck the glass and was knocked cold. When he came to he hastened to his club, where an attendant bandaged his head and gave him a stiff drink. "What the hell happened?" asked a sea gull. "I was flying across a meadow when all of a sudden the air crystallized on me," said the goldfinch. The sea gull and a hawk and an eagle all laughed heartily. A swallow listened gravely. "For fifteen years, fledgling and bird, I've flown this country," said the eagle, "and I assure you there is no such thing as air crystallizing. Water, yes; air, no." "You were probably struck by a hailstone," the hawk told the goldfinch. "Or he may have had a stroke," said the sea gull. "What do you think, swallow?" "Why, I—I think maybe the air crystallized on him," said the swallow. The large birds laughed so loudly that the goldfinch became annoyed and bet them each a dozen worms that they couldn't follow the course he had flown across the field without encountering the hardened atmosphere. They all took his bet; the swallow went along to watch. The sea gull, the eagle, and the hawk decided to fly together over the route the goldfinch indicated. "You come, too," they said to the swallow. "I—I—well, no," said the swallow. "I don't think I will." So the three large birds took off together and they hit the glass together and they were all knocked cold.

Moral: He who hesitates is sometimes saved.

James Thurber

The New Yorker, October 1939

Foreword

To all bird-lovers, it is a sickening noise – that loud thump on a windowpane, sometimes accompanied by a smear of bird snot or blood and/or a tiny feather. According to the author of this book, now widely regarded as the world's expert on window strikes by birds, we are losing per year between 100 million to 1 billion birds in the U.S. and 16 to 42 million in Canada to collisions with an unseen enemy. And because window collisions basically "set the table" for the millions of free-ranging pet and feral cats looking for easy prey to capture, and thus, biasing the staggering mortality figures associated with them, glass may well exceed our feline friends as being the highest form of human-related mortality for birds!

In recent years, the author has documented that 1,348 (12.8%) of the approximately 10,500 bird species in the world, and closer to home, 267 (29.7%) of the 898 species in North America, are known to be window strike casualties. And the worst news is that, despite its sophistication, the avian eye is simply not capable of detecting the difference between unobstructed air space and clear glass. It is an indiscriminate killer, killing fit or unfit, common or rare, big or small, male or female, or young or old birds. Victims are usually startled by loud noises, passing cars or arrival of a larger bird at the feeder, while others are involved in tail chases with their own kind or escaping the talons of hawks.

Some are killed instantly or knocked unconscious only to die later either from their injuries or from scavengers, some recover to fly off weakly, and the lucky ones are just shaken up but unaffected. One should not always assume that if a bird recovers in one's hand and flies off, then all is okay. The damage to the inner ear structures, the brain, neck muscles, and/or nervous system can be very debilitating and in many cases, deadly in the long term. If you find this hard to believe, take a run at a thick plate glass window without protective head gear and you will understand! On second thought, take my word for it.

Overall, the author has determined that birds are particularly vulnerable to clear or reflective large (> 2 m^2) windows at ground level and at heights above 3 meters. It doesn't matter whether the windows face north or south during migration times. Especially bad are glass corridors, stairways or rooms that create the illusion of clear passage. Most people do not realize that placing large plants inside glass-enclosed lobbies or expansive living spaces creates a false habitat that regrettably attracts unwary birds.

There are two somewhat separate issues involving window strikes – collisions with windows on city buildings and birds striking windows on residential buildings. Hopefully, business owners, building managers, architects, and landscape planners, now more cognizant of the problem, are taking the necessary steps to minimize window strikes in downtown cityscapes by using special fritted glass, avoiding mirror buildings, and turning off lights that attract migrating birds at night that in turn expose them to the window illusions that result in their death. But what can you do to minimize bird collisions around your home or apartment or your cottage?

First, one doesn't have to shut down a backyard feeder operation to stop birds from hitting your windows. Just use a little common sense. Place your feeders within a meter of any window surface to minimize any momentum a bird might gain when startled; close feeder placement eliminates the window threat. Alternatively, considering feeder placement at greater distances, study the landscaping around your windows and then put your feeding stations in places that guide visitors away from the glass; your goal should be to direct the coming and going of birds to the feeder in such a way that they move away not toward your windows. As for deterrents, here's what is worth trying and what is useless, according to the author. Forget using single wind chimes, blinking lights, hanging plants, large eye patterns, falcon silhouettes, or owl decoys. Uniformly covering windows with these very objects or others like cloth or silver mylar strips on or near the glass surface and separated by 5 to 10 centimeters will transform windows into barriers that birds will see and avoid. Even keeping white cloth drapes or sheer curtains closed during daylight hours work well, but only when the sun is shining through the window to make these coverings visible and permit birds to see and avoid the glass they cover. At other times, without the sun falling directly on these window coverings, the outside surface of the window reflects the facing habitat and sky, and projects a deadly illusion of unobstructed space. For serious "problem" windows, install insect screening or black plastic garden-protection netting mounted on frames away from the window. A more permanent solution entails adding strips of bird-friendly adhesive tape about 10 centimeters apart; see http://abcbirds.org/program/glass-collisions/ and http://flap.org for details about this and other cost-effective bird-window deterrent methods. As I always like to quip, let's remove the pain from our windowpanes for our feathered friends!

But don't take it from me! If you are reading these words, then you have right here in your hands the very means to learn much, much more about the subject from no less than the world expert on the subject – Daniel Klem Jr. himself!

Dan and I have been friends for three decades or more. While I cannot say that we are so close as to be weekend warriors sharing a beer or two on a frequent basis, we do know each other well enough to hold a solid mutual respect for one another and to share a love for birds. He has lived, breathed and studied the subject of bird collisions with windows for virtually his entire career, and he is a big believer in public outreach about wildlife conservation. I have happily served as one of Dan's disciples spreading the word about the subject for all those years.

As you will read in the Prologue, Dan first obtained his B.A. at Wilkes University in Wilkes-Barre, Pennsylvania and his M.A. at Hofstra University in Long Island, New York. After a brief stint in the military, Dan not only survived the horrors of the Vietnam war but also earned a Bronze Star Medal. And, as so often seen in many life stories of students starting out in the world of academics, he was serendipitously inspired by his interactions with Dr. William G. George of Southern Illinois University at Carbondale to undertake his doctoral studies on window collisions by birds. Dan made good use of his degrees and he currently serves as the Sarkis Acopian Professor of Ornithology and Conservation Biology at Muhlenberg College in Allentown, Pennsylvania. According to Wikipedia, his research has actually influenced building architecture, and he holds several patents related to window design. He also received an honorary doctorate from Wilkes University, where he is also a trustee.

Through his meticulous research and study trials over his entire career, Dan rose to fame in the worlds of both the ornithological and wildlife conservation communities as the reigning expert on bird collisions with glass and how to prevent them. Just like those biologists who choose to study cat predation on birds and find themselves as the targets of vitriolic rants and even worse behavior by misguided folks who simply and selfishly do not want to face the truth, Dan was thrust into that very same sphere. You see, no matter where they live in the world, even today, most people just do not want to know that the large expanses of glass in the windows of their homes and workplaces kill a heckuva lot of birds – and needlessly so. Sadly, these same folks will probably not read, let alone purchase a copy of this book. But fortunately for the billions of potential feathered victims in the future, other biologists, conservationists, and even better, media experts will pay attention to the wise words penned in this comprehensive very first volume published on this subject.

On a personal note on Dan's character, I do not think that I have ever met a more humble, modest and gracious person in the field of biology throughout my entire professional life. During a phone conservation with him while I was preparing this Foreword, Dan confessed to me that I had apparently helped him immensely in his early career by being supportive of both him and his cause, something I had long forgotten with these aging gray cells of mine. But you know something, folks, I am glad that I did. And so are the birds.

---- **David M. Bird, Emeritus Professor of Wildlife Biology, McGill University**

Preface

THE PURPOSE OF THIS BOOK is to offer evidence that will encourage you to join me in protecting the Earth's bird life from the lethal effects of windows. We humans are killing birds the world over when they fly into what is to them an invisible barrier: the windows in the buildings we construct. The victims are always innocent, have no means to protect or speak for themselves, and are never purposefully intended targets. Few if any other living creations provide humans with more value. For many, birds are the most prominent representatives of nature and human culture because of their aesthetic beauty, usefulness, service as symbols of our cherished and defining beliefs, and their association with good feelings and health. Most of these difficult-to-quantify contributions are literally priceless. We humans have the power to stop the senseless death of billions of birds, and your help is needed to do so.

The task of making our windows safe for birds is more likely to be successful when like-minded individuals combine their passions, talents, and energies. There is the promise of strength in numbers when informed individuals take meaningful action. Working together, we can limit and ideally eliminate these tragic casualties.

The window hazard issue has been slowly attracting attention among the bird conservation community for decades. Compared to earlier decades, a remarkable degree of awareness and action has appeared in the past few years. Increased education, especially through media coverage, the development of bird-safe manufactured products, and use of bird-safe architectural design and implementation are all signs of hope around the world. And yet, these changes are too few and as yet not meaningfully effective. We need much more from many more to effectively address the senseless and harmful killing.

Reasoning individuals that have a view of the world beyond themselves or their species have a responsibility to learn about this too often ignored and under-appreciated loss of life: its scale, the consequences for specific species and birds in general, and ways to mitigate or ideally eliminate this source of human-associated avian mortality. To provide information and rationale, I document in this book the results of my research over 47 years, plus that of others, from the earliest window-kill accounts to the first comprehensive investigation in the early 1970s and thereafter. This book attempts to define the scope of the problem, solicit change, and guide the steps needed to solve it.

My first thought was to write as impersonally and objectively as possible, simply to record the overwhelming evidence that windows are a serious global threat to birds and describe how to deal with this threat. I soon realized, however, that my personal journey in studying this topic could also be enlightening and potentially helpful to this life-saving cause. Too often, that journey has been punctuated by my failure to meaningfully educate and gain the help of those in positions to do better. By including anecdotes about some of my personal setbacks and achievements, I intended to provide an entertaining human-interest component in a book that otherwise would be a very grim read. To be sure, this is a serious topic requiring serious attention, but at times humor has its place. My wish is that you occasionally find something herein that makes you smile, if not laugh out loud, if only to relieve the stress of dealing with an issue that for birds is a matter of life and death.

Prologue

How did I get here—sitting on a bench just before dawn, waiting to see if birds would strike nearby windows? The answer begins where I began. I grew up in northeastern Pennsylvania, and one of my earliest recollections is of enjoying the outdoor tales of my Russian-speaking grandmother. Her stories came from her youth at the end of the 19th century in the Carpathian Mountains of Galicia, a region that spans southeastern Poland and southwestern Ukraine today. As a young girl, she was regularly in the field watching and attempting to protect their sole cow and other family livestock. Wolves, a threat to humans and their possessions, dominated her stories, and for me, then and now, her tales of these charismatic predators were spellbinding. More specifically, my father and his brothers, primarily my Uncle Nick and his children, stimulated and shaped my interest in wildlife. My fascination with birds began with my first opportunity to roam the fields and forests alone, but most often with my cousin Connie and brother Dave, who were both a bit over a year my junior. Every living part of the field and forest, especially birds, caught our attention, and fueled our interest and imagination.

My parents expected my two brothers and me to attend college, but their ability to help us was limited. Their efforts consisted of encouraging us to seek the counsel of our teachers and use the resources of our local library. I questioned my academic competence and had no guiding mentor to offer encouragement. My weak academic and social confidence made it a struggle to get into and through college, but I managed it at a local school, Wilkes College (now Wilkes University), then and still dedicated to educating first-generation college students, of which I was one.

I was a commuter, living close enough to the college to walk to the campus. I majored in biology. As is true at many liberal arts colleges, most biology majors were interested in health science careers, especially aspiring to practice medicine. My college performance, heavy on both trial and error, was not good practice for success. Other than a career in the health sciences, I had little understanding of what else one could do with a degree in biology. Very early on, if I had flirted at all with the idea of health science opportunities, those notions were quickly dispelled by my performance. My future direction needed more investigation, and putting my heart, mind, and persistence into the task, I discovered the prospect of studying and eventually finding a career working outdoors. My love at first sight with the Atlantic Ocean and my fondness for fishing prompted me to consider advanced training in oceanography.

To my surprise, amazement, and pleasure, I was accepted into a graduate marine science program during my senior year, in the spring of 1968. At the same time, our country was in violent turmoil over the Vietnam War, and soon I was to discover how it would affect me. The choices became simpler but also more difficult and frightening. President Lyndon B. Johnson revoked student exemption from the draft for graduate study. The use of draft numbers that subjected conscription to a lottery would come later. With my upcoming graduation, my draft status changed from S-2 to A-1. Contemplating my options, which included an imminent conscription summons or enlisting, I chose military service, which promised me some choice in what service I would be performing. Most of my classmates in the same situation chose various means to avoid military service. Their reasoning was understandable but for me did not feel justified, based on my gratitude for what our family of eastern European immigrants understood as the benefits of citizenship—freedom of choice, education and other opportunities—and our obligation to the government that provided them.

If permitted, I surely would have continued my studies to fulfill my dream of becoming a scientist. Just before I learned about my draft prospects, I had prepared 15 graduate school applications. One had already been mailed. The others went into the trash. The sent letter resulted in a surprise acceptance to Northeastern University in Boston, Massachusetts. After informing them about my military status and induction, they agreed to honor my acceptance upon my discharge from the armed forces. I fully expected to accept their offer. For now, I was in the U.S. Army by July, continuing until the spring of 1971.

About a year before returning to civilian life, my plans changed again. This time from a developing distant relationship with the woman who became my life partner. We met while I was home on leave and prior to my being assigned to serve in the Republic of Viet Nam. She lived on Long Island, New York, and not Boston. Soon I was married, enrolled and in residence at Hofstra University in Hempstead, New York, in the fall of 1971. I still planned to study marine science, but the subject and faculty practices were not welcoming at Hofstra.

Fortunately for me, the Department of Biology had an enthusiastic professor of ornithology who was eager to attract and accept students. That professor was Paul A. Buckley. He invited me to study birds with him. For the first time, I enjoyed the academic and personal attention of a mentor who met my best expectations and fulfilled my desire to study outdoors. My subject of study would be birds, not fish, but I could not have been more excited and pleased. I eagerly seized the opportunity, and doing so has enriched my personal and professional life ever since.

William G. George, ornithologist and author's doctoral advisor in the Department of Zoology, Southern Illinois University at Carbondale, with Hairy Woodpecker, circa 1970.

In my last year at Hofstra, I applied to a doctoral program in behavioral ecology within the Cooperative Wildlife Research Unit at Southern Illinois University at Carbondale (SIU-C). Soon after arriving, my advisor, Willard D. Klimstra, informed me that I would be expected to study game birds unless I could obtain my own funding. I found all birds interesting and worthy of study but especially aspired to conduct a behavioral study on birds of prey. Specifically, my grand idea was to compare prey-catching methods across species in hopes of validating existing interpretations or revealing new evolutionary relationships. Just a year after my arrival, the political climate changed within the Department of Zoology, where I was officially enrolled, and these changes forced me to consider searching for a different graduate school opportunity. With no meaningful connections, I had no realistic prospects. Aware of my predicament, the department's ornithologist, William G. George, invited me to study with him. We knew each other through my serving as his teaching assistant in two undergraduate courses, Vertebrate Zoology and Ornithology.

We discussed a few possible dissertation research topics, to include my ideas on birds of prey. During this first meeting, he was agitated by a student encounter, and it prompted him to tell me about birds and windows. The student was the son of a faculty member and chair of the Department of Religion. He was an especially good birder, and Dr. George asked him to collect an unusual nest he had discovered on their family property. The student refused to take the nest, explaining that he believed it was unethical to do so. With visible irritation, Dr. George told me that the family of the student lived in a modern new home lavishly covered with glass. Some 50 to 60 birds annually were killed flying into their windows. He railed about this, with consternation that at no time had they ever described these deaths being unethical when offering the dead birds to him for inclusion in our bird museum. He then recommended I consider looking into studying collisions between birds and windows as another potential dissertation research topic. I was intrigued, but, like most at the time and many still, I had not heard about birds flying into windows.

The very next morning, in January 1974 as best I can recall, I sat on a bench in front of the glass wall of the Neckers chemistry building, where Dr. George informed me window strike casualties had been collected and turned in to him. I arrived in the dark about 5 a.m. Over the next 30 to 40 minutes, the dawn light brightened enough to see the surroundings. Within minutes of my being able to see details, to my right, a swiftly flying bird flew through the leafless branches of a tall oak and into the upper story of the building facade. The bird appeared to be killed outright and fell to the base of the sheet glass wall. It was a Mourning Dove. I was stunned figuratively, but not literally and fatally, as the dove had been. After collecting the victim, I eagerly began a search beneath the windows of other campus buildings. This initial search revealed feather and skeletal remnants at several sites. I presumed the remains were other window-killed casualties, given their location beneath windows, and feathers, feather imprints, and body smudges on the glass surface. The strike I witnessed and these first finds were compelling evidence suggesting a topic worth further study. I was hooked.

Distant and close-up views of the Neckers chemistry building on the campus of Southern Illinois University at Carbondale, at which the author recorded his first bird-window strike.

Chapter 1

Humans Make Glass, Birds Hit Glass

Glass windows have existed since as long ago as 290 CE, if only in a limited supply of small sheets. It seems fair to say that window glass has enriched human aesthetic, cultural, physiological, and psychological well-being for at least 16 centuries. Even one small pane is enough to admit a bit of the sun's light and warmth into an enclosed space. The tendency of builders, and the willingness of their clients, to use this product in large quantities apparently resulted from the need of human society to seek safety within the solid walls of dwellings away from the reach of marauders. Sheet glass permitted viewing the out-of-doors from the comfort and protection of indoors. In the Middle Ages, ecclesiastical interests led to the lavish use of both tinted and clear panes. These windows were used in the cathedrals of Europe and then in the domestic dwellings of the rich, especially in Tudor England. The technical ability to manufacture large sheets of glass was developed at the turn of the 20th century. With the building boom that followed World War II, 1945 to the present, flat glass has become a prominent, even dominating, construction material used in the majority of human dwellings and other structures. In 2009, 6.6 billion m^2 (6,600 km^2 = 2,548 mi^2) of flat glass was manufactured worldwide, about the area of the U.S. state of Delaware, at a value of $23.54 billion. The amount of glass used in construction has continued to increase annually.

The history of window glass as a source of bird fatalities is similarly ancient and progressive. The confirming obituaries, however, do not begin to appear in the literature until well after 1800, with the development of modern ornithology in Europe and North America. Thomas Nuttall published the first scientifically documented window fatality in his 1832 *A Manual of the Ornithology of the United States and Canada*. He described how a hawk in pursuit of prey flew through two panes of greenhouse glass only to be stopped by a third. The next account was by Spencer F. Baird and his colleagues Thomas M. Brewer and Robert Ridgeway in volume one of their 1874 three-volume work *A History of North American Birds*. They described how a shrike struck the outside of a clear pane while attempting to reach a caged canary.

Cartoons are copied from Bird-window Collision Presentation Toolkit, Version April 10, 2019, with permission from Bird-window Collision Working Group (BCWG).

Three decades passed before the next detailed bird-window collision report. Charles W. Townsend, writing in the fourth issue of the 1931 ornithological journal *Auk*, published the first account of a series of fatalities for a single species. This was the first suggestion that avian vulnerability to windows may be more marked in some species than in others, and that a single window may claim a succession of victims. His account suggested that he had records of a number of window-killed species. The following quote highlights what he believed was a particularly vulnerable one: "The Yellow-billed Cuckoo seems especially prone to run its head against windows if I may judge by five instances that have come to my attention. The first one to do so, whose skin is still in my collection, killed itself on June 13, 1876, by flying against a window in Jamaica Plains, Massachusetts, as did also another at the same window on June 9, 1878. A third committed suicide on June 10, 1904, against a window of Mr. William Brewster's museum in Cambridge." These records are historically interesting in that Townsend's language describing window strike victims as "tragedies" suggests that he regarded them as rare, self-destroying incompetents, or "cuckoos" in the human sense of the word. As noted, sheet glass was relatively rare at the turn of and early 20th century, and it is likely there seemed to be little reason for concern about the relationship between windows and birds. Today, with the ever-growing presence of sheet glass in the human-built environment, it is common to find modern buildings that are entirely surfaced with glass, and the fatal consequences for birds are increasingly imposing.

From 1974 through 1979, I found 88 reports in books and journals from North America, South America, the West Indies, Europe, and Africa that described birds striking windows and documented species-specific accounts. Most were from the United States, but others were from Canada, England, Germany, Luxemburg, Norway, Netherlands, Rhodesia, and Switzerland. Remarkably, textbooks and encyclopedia treatments about birds presented little, if any, description of the fatal hazards that windows pose to birds. The sheet glass industry and its commercial allies to this point appeared to be unaware of the problem. By contrast, my personal inquiries to individuals repeatedly revealed that bird strikes at windows were common knowledge among those who were attentive, even with only a modest familiarity with birds.

From these early recordings to the present, scientific research papers and accompanying popular articles have continued to grow. Current publications document extensive details that include: (1) quantitative studies revealing species, building, and environmental conditions; (2) individual injuries and causes of death among strike victims; (3) the level and composition of mortality as a species-specific conservation concern; (4) means to prevent bird-window collisions; and (5) a number of reviews updating the latest body of collective knowledge. This growing body of literature has resulted in increased awareness and action among conservation-minded constituencies that now include saving birds from windows as part of their mission.

From the very beginning, the evidence I compiled and subsequent reinforcing and validating study results of others made it clear that the killing was taking place at the windows of commercial, residential, and institutional buildings in urban, suburban, and rural settings. Hundreds of birds a day kill themselves flying into the first few stories of prominent glass-covered high-rise buildings in cities during migratory periods. These dramatic accounts, documented by organized volunteers, often appear in newspapers and have attracted national and international publicity. Far more dramatically, several reinforcing investigations reveal that most victims, hundreds of millions of common birds as well as species of conservation concern, are dying by striking clear and reflective windows in every season, under all kinds of weather conditions, at panes of all sizes, by the ones and twos at houses throughout human-occupied lands. Most of the total deaths in any calendar year result from hitting sheet glass in various sizes of residential and small commercial buildings, and school buildings of all types. Study after study documents and validates the claim that lethal collisions occur wherever birds and windows coexist.

Distant and close-up images of a window-killed female Northern Cardinal. Photographs by Peter G. Saenger.

Yet most of the conservation community is still to be convinced that the sheet glass threat to birds is real, substantial, and merits their commitment and action; this includes educators, scientists, citizen-scientists, and wildlife law-enforcement. Members of the building industry—glass manufacturers and fabricators, structural and landscape architects, developers, and building managers—require persuading that there is a market for creating and employing bird-safe windows in residential and commercial structures and in noise barriers along highways and railways. Among the diverse public, there are those who will be moved to make windows safe for birds because they are mortified at the thought of these attractive animals suffering and dying from a preventable cause. These people need no convincing. Others accept that suffering and death is part of life, and though saddened to see a healthy individual cut down in its prime, they need convincing that this source of avian mortality is actually reducing and threatening the survival of species populations. A third category of people are those who do not care, no matter what, for a number of unpleasant and irrational reasons. These people cannot be convinced. This book is written for those who can be convinced.

The primary purpose of this book is to stop the unintended and unwanted killing of the defenseless innocent, those that have no voice or other means to protect themselves from an invisible killer. I hope to enlist the power of citizens everywhere to help solve this problem. The goal is to inform, persuade, and incite a change in human behavior to produce a life-saving change in bird behavior.

Window-killed adult Sharp-shinned Hawk.
Photograph by Peter G. Saenger.

The public needs information to act meaningfully to stop this senseless loss of attractive and useful life. However you view yourself and your place in society, your help is desperately needed to stop the slaughter. We know enough already about how to stop it. Our collective efforts will not be comprehended by those we seek to protect and save. But without acting to save the birds that provide us with recreational joy, at minimum, we weaken and damage our hopes for a healthy world to pass onto our children. We even damage our economy. Making windows safe for birds will reward all humans, whether they care about birds or not, with their continued survival and the many benefits they provide.

Chapter 2

Why Birds Hit Windows

Gordon Lynn Wall's classic 1942 book *The Vertebrate Eye and Its Adaptive Radiation* describes the avian eye and visual system as superior to those of all other backboned animals. This assessment remains valid, but as exceptionally gifted as the avian visual system is, it is reasonable to infer that birds strike windows because they fail to recognize them as obstacles. The only alternative explanation is that birds see clear and reflective windows and fly into them for the purpose of killing or injuring themselves. This type of self-destruction and injury is contrary to individual and species survival. Extensive evidence suggests the conclusion that certain windows are simply functionally invisible to birds under certain conditions.

Various ideas have been offered to explain how windows are rendered functionally invisible to birds. One hypothesis presumes that most or all birds are able to see windows, but those that fail to do so are individuals whose visual systems are defective, impaired in some manner, or deceived by one or a combination of window-associated environmental features. The other hypothesis presumes that the visual system of birds is incapable of perceiving sheet glass: immature birds because their vision is not fully developed, immature or adult birds because they lack some learning component that might enable them to detect and avoid glass. Some would modify that hypothesis to say all individuals of all species are simply incapable of perceiving transparent and reflective windows because of their unique properties and the anatomical and physiological limitations of the avian visual system.

In 1945, George Willet, writing in *Condor*, speculated from several window strike reports in southern California that migrating Swainson's Thrushes hit windows because of some physiological deficiency affecting their eyesight. His interpretation cannot be proved wrong and may be true in some rare cases, but this potential cause is unlikely to explain most strikes, given the diversity of victims. Swainson's Thrush is no more or less vulnerable to windows than several other species.

Authors of several reports explain window strikes resulting from impaired vision. The range of impairments might include the effects of alcohol, distractions that divert attention, smoke, blinding glare, and weather conditions that hamper visibility.

Bird literature contains accounts of waxwings, robins, and sapsuckers becoming intoxicated on fermented foods in the wild. In 1978, it was Steven Rogers writing in the *Colorado Field Ornithology (C.F.O.) Journal* who specifically suggested that birds may strike windows because they are under the influence of alcohol. He described the unpublished work of Walley, who found that window-killed waxwings in Dauphin, Manitoba, had enough alcohol in their blood to be classified as legally drunk. As interesting as this account is, the majority of species documented striking windows are not known to have consumed alcohol. Alcohol consumption to explain most window strikes seems unreasonable. But, for those birds that eat alcoholic foods to the point of impairment, it is reasonable to conclude those that "drink" and fly are apt to be more vulnerable to windows and other mishaps than those that do not. Most birds that strike windows do not behave as if intoxicated, and almost certainly are not drunk.

There are several accounts of birds hitting windows while being pursued by predators. The interpretation is that the victim was distracted and not paying attention to the window as a barrier. My records contain many accounts of both pursued and pursuer being killed striking a window, but the majority of these incidents seem completely unaffected by any external distractions.

In 1960, H. P. Langridge, writing in the *Florida Naturalist*, speculated that smoke was responsible for impairing the senses of several hundred migrant warblers that hit the windows in a shopping center over a four-day period in the Palm Beach area of Florida. The source of the smoke was burning trees, bushes, and other vegetation a few miles to the south. Smoke was implicated because a similar event occurred under similar conditions in 1945. To my knowledge, there are no other accounts of smoke causing bird-window collisions, although it cannot be dismissed in these two incidents.

Without supporting evidence, in 1972, Paul Sinner provided an unpublished account from Valley City State College in North Dakota suggesting that some birds may strike windows because they are blinded by the sun's glare off the surface of the pane. His explanation may be true, but if so it is probably a very rare event. Alternatively, my records contain many accounts of birds hitting windows under overcast conditions when sun glare did not occur to the human observer. Similarly, poor visibility due to weather conditions was inferred as a cause of window collisions in 1963 by Claus Konig, writing in the German journal *Deutsche Sektion des Internationalem Rates für Vogelschutz, Bericht*. He describes birds being more vulnerable to windows on misty days. My records do contain one uniquely dramatic account of 50 Dark-eyed Juncos hitting a window in a rural southern Illinois house in a snowstorm. Most accounts of birds striking windows occur during favorable weather, with unhampered visibility.

Detailed analyses of accounts reveal that no one or a combination of habitat or human-made structural features are solely responsible for birds striking windows. The most reasonable universal explanation for why apparently healthy wild birds fly into and kill themselves striking windows is the suspicion that the anatomical and physiological limitations of the avian visual system do not permit birds to see clear and reflective panes. Supporting this interpretation is the sheer diversity of victims and the characteristics of collision sites.

Indirect evidence supporting the claim that sheet glass is invisible to birds is in the methods of the now famous biological and psychological studies by Eleanor Gibson and Richard Walk writing in *Scientific American* in 1960; Walk and Gibson writing in *Psychological Monograph* in 1961; John Emlen, Jr. writing in *Behaviour* in 1963; and R. B. Tallarico and W. M. Farrel writing in the *Journal of Comparative Physiological Psychology* in 1964. They conducted so-called "visual cliffs" experiments. As part of the methodology of these studies, the authors accepted, and their results confirm, that birds and the other vertebrate animals tested were unable to see clear glass or plastic sheets as a supporting surface. The respective subjects acted as if it was invisible in the experimental set up.

The anatomy and physiology of the avian visual system offers direct evidence for the interpretation that windows are invisible to all birds. This claim is justified from evaluating the results of investigators studying the structure and function of the bird eye. Among others, the avian vision science of Robert Beason, Mark von Campenhausen, Timothy Goldsmith, Olle Hastad, William Hodos, Kuno Kirschfeld, Graham Martin, Anders Odeen, and George Walls are especially relevant and noteworthy.

Birds have eyes structured like but larger than any other backboned animals in relation to their overall body size. An essential avian perception principle is "We are not them." Birds are not capable of telling us what they see, at least not directly. The best humans can do is *interpret* what we think birds see, based on their physical and functional accoutrements and their behavior.

Like humans, birds see using their eyes and their brain. The outside of the eye is covered by a clear cornea, behind which is a clear fluid. Then further inward is a clear lens, followed by more clear fluid. Still further, at the very back of the eye orbit, is the retina, consisting of complex layers of specialized cells. In a healthy eye, the combination of retinal and other neural cells transforms electromagnetic wave energy into electrical energy that is sent on to the brain. The visual centers of the brain receive varying input signals resulting in what we and birds see. For sure, it is the brains of birds and people that do the seeing, not the eyes.

With all that humans share in common with birds, there are big differences between what we and they see. At least, that is what we interpret, based on our respective anatomical and physiological differences. These differences suggest that in general birds see a clearer, larger, and more colorful world than we do. They have more muscles associated with more eye elements, permitting better focusing, formally known as accommodation. The cells receiving the light energy that enters their eyes are more concentrated in some parts of the retina than ours. The areas of greater cell density are called *fovea*. Whereas humans have a single fovea in the center of the retina, birds have two or more. We judge that this difference offers them greater acuity over a large visual field, if not their entire one. Like us, birds have rod and cone cells that are stimulated by the wavelengths of light that enter our eyes. Light first passes through the cornea, the lens, and fluids on each side of the lens. The rod cells are as much as a hundred times more sensitive than cone cells. With their single visual pigment, rods inform the brain about intensity, or strength of dark to light. Cone cells have distinct pigments adapted to absorb select wavelengths of light. Cones transform light energy into electrical signals that are sent to the brain and perceived as color. As a component of vision, color is a brain function, just like the image overall that we and birds see.

Human color vision is described as trichromatic. We possess three different cone cells with visual pigments that have three different absorption peaks, generally described as blue, green, and red. The range of wavelengths perceived by humans is from 400 to 700 nanometers (nm = 10^{-9} = 0.000000001 m, or one billionth of a meter). The respective absorption peaks are 424 nm (blue), 530 nm (green), and 560 nm (red). The human brain sees white when the three cone types are stimulated to their maximum. The brain interprets intermediate or blended colors such as turquoise or magenta when the other cone cells are not stimulated to their peaks.

Birds have tetrachoromatic vision. This is interpreted as an enhancement compared to what we see. Instead of three types of cones, they have four. Three have peak absorption of blue, green, and red, and the fourth absorbs wavelengths in the ultraviolet area (UV, 300-400 nm). We humans cannot see UV. Our lens absorbs it before reaching our retina. In the avian eye, UV passes through the lens to activate a pigment in a distinct retinal cone cell. Birds see white when all four cone types are stimulated to their maximum. Avian physiologists refer to what birds see as bird white to differentiate it from what we see as white. We have no accurate understanding about what *bird white* looks like to birds.

Birds have two structural types of cone cells, and this is another differentiating feature thought to enhance bird vision in comparison to ours. Bird cone cells also contain oil droplets, which are endowed with special light-absorbing pigments. We humans have only single cones. Birds possess single and double cones. The functional difference between the two is still largely a subject of investigation. The function of oil droplets, however, is better known. The pigments in oil droplets filter out select wavelengths of light. This narrows the range of wavelength absorbed by the different visual pigments in the different cone types.

The ability of birds to see UV is predicted to hold potential for protecting birds from windows. The premise is that UV signals applied to windows will transform them into obstacles that birds will see and avoid. Since humans cannot see UV, we would continue to enjoy the unobstructed views that we expect and value from windows. The role of UV in preventing bird strikes is a very important topic.

Unless marked with visible dirt, soap, or artificial elements, clear and reflective glass is not perceived by birds, humans, or any other backboned animals, based on what we collectively have learned about the makeup and function of their respective visual systems.

Chapter 3

How Big a Problem

To put the scale of losses by various human-related bird killers in perspective, my speaking and writing, starting in the late 1980s, included "sound bite" tactics. The purpose was to capture the attention of anyone who might be moved to listen and take action. One such tactic involves comparing bird losses at windows to those from higher-visibility oil spills. Prominent oil spills that have captured international attention as environmental disasters include the Exxon Valdez disaster and the more recent Deepwater Horizon fire and spill in the Gulf of Mexico. By any assessment, the Exxon Valdez oil spill was a horrific environmental disaster. The oil tanker Exxon Valdez released 260,000 barrels of crude oil into Alaska's Prince William Sound on March 24, 1989. The spill was estimated to have killed 100,000 to 300,000 marine birds. The 2010 Deepwater Horizon spill is estimated to have caused 1,000,000 bird deaths. By contrast, in the 1970s my original and lowest estimate of losses from window collisions in the U.S. alone was 100 million bird kills a year. This minimum window-kill number equals the toll from 333 annual Exxon Valdez disasters and 100 Deepwater Horizon oil spills every year. Yet those writing in the media about assaults on the Earth's environment are either unaware, unconvinced, or willing to overlook the horrific loss of bird life occurring at windows.

Currently, assessment of reliable available data suggests free-ranging cats kill more birds than any other human-related avian mortality factor. This claim is based on a sophisticated quantitative analysis of the most objective evidence. The toll was published in 2013 for Canada by Peter Blancher of Environment Canada and the Canadian Wildlife Service, in the journal *Avian Conservation and Ecology*, and by Scott R. Loss and his colleagues principally affiliated with the Migratory Bird Center, Smithsonian Conservation Biology Institute, in the journal *Nature Communications*. They, respectively, estimated annual cat kills at 100 million to 350 million in Canada and 1.4 billion to 3.7 billion in the

U.S. Other works similarly examining birds killed striking windows were published by Craig S. Machtans and his colleagues principally affiliated with Environment Canada and the Canadian Wildlife Service, also in the journal *Avian Conservation and Ecology* in 2013, and by Scott R. Loss and his colleagues in the journal *Condor: Ornithological Applications* in 2014. They, respectively, report 16 million to 42 million for Canada and 365 million to 988 million for the U.S. The figures attributable to cats, however, as these authors acknowledge, are compromised in that cats are known to regularly patrol and take dead and injured birds after those birds have first struck a window. Among a large list of threats, bird kills from select other human-associated sources include 120 million from hunting, 6.8 million from striking communication towers, 60 million from vehicle road-kill, and tens of thousands to millions from wind turbine strikes, power line strikes and electrocutions, pesticides, and other poisons and pollutants.

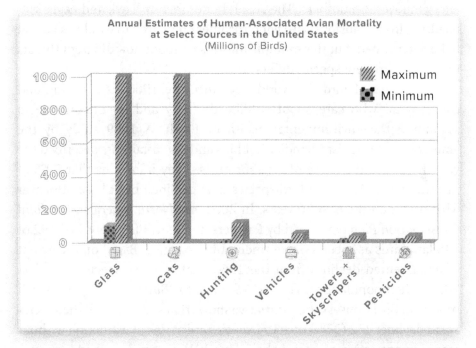

Comparing all the human threats to birds, few, if anyone, would argue that the most important is habitat destruction. Destroy the habitat upon which life depends for food, shelter, and reproduction, and you destroy the ability of any life to survive. To be sure, to save select species from unrecoverable debilitation and extinction, we need to set priorities regarding which species need our attention first or most.

Notwithstanding these immediate needs, it should not matter what the ranking is, based on the level of loss from human-caused avian mortality. With the exception of a few select managed game species, the mortality of free-living wild species populations attributable to humans is harmful. We need to address all types of unintended harm that humans exact on species with which we share the planet, to at least mitigate these harms and ideally eliminate them.

Human threats carry varying levels of consequences for the population health of particular species. For example, cats and windows may result in far more bird deaths than wind turbines, but wind turbines may kill a disproportionate number of hawks and eagles. Hawk and eagle populations are likely disproportionately harmed by fatalities that occur at wind turbines because of their longer lifespan, smaller populations, and larger home ranges than the perching bird species that dominate cat and window kill numbers. The absolute number of hawk and eagle kills resulting from collisions with wind turbines compared to windows has yet to be determined, but this example informs us about how different threats may affect different species differently.

The loss of bird life worldwide since the 1960s is alarming and concerns all who care about wildlife diversity and its importance in sustaining the environmental health of Earth. A 1989 study by the eminent ornithologist Chandler Robbins and his co-authors described the vulnerability and decline of migrant birds. They inform us that 50% of migrant North American bird species have declined by at least 50% over what was then the past 50 years. In 2004, the *North American Landbird Conservation Plan* developed by Partners in Flight (PIF) identified 192 of 448 landbirds as species of continental importance. Based on the extent of documented declines from the 1960s, 29 of these species have lost 50% of their populations. The *2016 State of the Birds* report by the North American Bird Conservation Initiative shows that well over half the species are in decline; 432 (37%) of 1,154 species are at risk of extinction without management action on their behalf. This 2016 report described how one-third of North American bird species need urgent conservation action. Although no one is certain of the accurate number, a general estimate suggests 10 billion individual birds are present in North America at the beginning of the spring breeding season each year. That estimate increases to 20 billion individuals in the fall, when the young of that year have entered and increased each species' population numbers. In 2015, Scott Loss and his colleagues, writing in *Annual Review of Ecology, Evolution, and Systematics,* estimated that the annual toll exacted by window collisions reduced the overall North American bird population by 2% to 9%. The most recent assessment of human-related bird declines in North America that includes the substantive attrition from window collisions was reported in 2019 by Kenneth V. Rosenberg and his colleagues, writing in the journal *Science.* They document a net loss of three billion individual birds, or 29% of the 1970 abundance. Collectively, these are alarming figures.

The general loss of bird life worldwide is just as worrying, and consequently so is the declining population health of individual species. The toll that sheet glass and plastic in the form of windows exacts on most individual species and bird populations overall is unknown. The exceptions are species of conservation concern—the endangered, threatened, and vulnerable species for which every unintended loss can be potentially critical. What we do know about the window threat, however, is enough to cause concern for the population health of common species as well as species of conservation concern.

For wild animals in general, most natural population mortality is what is called *compensatory* mortality, with a disproportionate attrition, for example, resulting from starvation or predation one year, while disease or unseasonable weather claim more the next. These types of losses are described as compensatory because populations can replace the individuals lost through greater breeding success in subsequent years and thereby compensate for the attrition. A mortality factor for which the species is not able to compensate is called *additive* mortality. It is additive because this type of mortality is in addition to the natural losses species are known to recover from, between one annual cycle or more, and the next.

In 2011, Todd Arnold and Robert Zink, writing in the journal *PLoS One,* described birds killed at communication towers and windows as a compensatory mortality. They compared the known collision losses of select species with their estimated total North American population numbers. They concluded that those species with higher fatalities also had high population numbers, and described these findings as supporting the idea that collisions are a compensatory population mortality factor, in their words, having "... no discernible effect on population trends of North American birds." Frank Gill and colleagues, in Gill's latest 2019 *Ornithology* textbook, offers the same interpretation, based on the large numbers of most North American landbirds that are also known to strike towers and windows.

The potential flaw in comparing these fatalities and overall population size, at least for the window threat as a compensatory loss, is treating this source of mortality as equivalent to natural losses. According to this argument, the loss of less fit individuals in respective species populations is compensated for by more fit individuals, who in effect make up for the deaths of the weak by producing more young.

An alternative view is that, given the fundamental reasons that birds die striking windows, the fittest individuals in species populations are just as vulnerable as the less fit. This claim is based on the anatomy and physiology of the avian visual system and how birds behave around clear and reflective windows. Evidence supports the interpretation that windows are simply invisible to all individuals and all species of birds. Notwithstanding the disproportionately large number of landbirds killed flying into windows, the major groups and their respective species known to be window-kills span the full range of species diversity: select examples, among many others, are waterfowl, pigeons, grouse, hummingbirds, birds of prey, and woodpeckers. The diversity of dead include young and old, males and females, breeding and non-breeding, and migrants and non-migrants. Fatalities are recorded wherever birds and windows coexist. At the end of a breeding season, it may be that large numbers of window-killed juvenile American Robins can be considered compensatory mortality, with the expectation that fit adults will produce more of them in the next breeding season. By contrast, the window-killed large numbers of adults migrating to the neotropics after the breeding season, which must be among the fittest, will not return to produce young to sustain the health of their species populations. Many of these represent the very species expected and claimed to be able to compensate, such as thrushes, wood warblers, vireos, and sparrows.

The very nature of the problem that sheet glass and plastic pose for birds identifies windows as an indiscriminate killer. It is this arbitrary loss of the fittest as well as the unfit members of species populations that make this type of mortality additive—and as such alarming because of the consequences of these losses to species health and survival. Thinking that common species as well as those of conservation concern can afford to lose the healthiest members of their respective populations indiscriminately is expecting too much. Surely, we need to address all the unintended causes of human-associated avian mortality until we can determine the actual consequences of each, as Loss and O'Connell have recommended in the 2018 *Ornithology* textbook by Morrison and colleagues.

The lethal threat windows pose to birds should be of special concern because of the extravagant amount of sheet glass present in human structures and its potential to be an indiscriminate and additive killer. Compared to cats as avian predators, the growing number of windows around the world predicts concomitant increases in collision victims. The documented carnage justifies attention and preventive action, whether window strike fatalities are first, second, or further down the tally of dead caused by or associated with the human-built environment. It can't be emphasized enough that, whereas avian species populations compensate for natural lethal threats and their resultant attrition, such as starvation, disease, and predation, which vary in severity, the invisible lethal threat that windows exact on birds is constant and consistent, an unrelenting addition to all the natural compensatory losses to a species, year in and year out.

Domestic cat tracks below a window where injured and dead birds occur after striking. Photo: Peter G. Saenger.

The bird species documented as window victims is highly diverse. The total to date is 1,348 (12.8%) of the approximately 10,500 bird species known worldwide. The records were compiled from my original survey of North American museums and select individuals, published accounts, and additional systematic surveys conducted by my co-investigators and students. The major groups of birds not well represented as victims are those not typically found around human dwellings. These include waterfowl, pelagic, and other bird life in habitats with few or no human structures. Window-strike casualties of conservation concern listed by the National Audubon Society in their 2007 WatchList for the U.S. are 6 (9%) of the 67 species on their Red List, and 24 (26%) of 94 species on their Yellow List. Red List species are declining rapidly and of global conservation concern. Yellow List species are declining but at a slower rate and are of national conservation concern. In their 2014 paper Loss and his colleagues, writing in the journal *Condor: Ornithological Applications*, listed the following species with declining populations to be especially vulnerable to windows in the U.S.: Golden-winged Warbler, Painted Bunting, Canada Warbler, Wood Thrush, Kentucky Warbler, and Worm-eating Warbler.

With very rare exceptions, such as the cause-and-effect linkage between the pesticide DDT and its role in the decline of select raptors and pelicans, the causes of diminishing common species as well as those of special conservation concern are complicated and mostly uncertain. Currently, one species suspected to be adversely affected by window-strike mortality at the population level is the Swift Parrot of Australia. This species is classified as world threatened. In 2006, Raymond Brereton, manager of the Swift Parrot Recovery Program for Parks and Wildlife Service of the State of Tasmania, informed me that 1.5% of the 1,000 breeding pair population annually succumb to window collisions. This attrition was so alarming, the Australian government implemented a window collision education program to encourage people to provide the species with protection.

Other documented window casualties and their respective international conservation level of concern include the following: Critically Endangered – Townsend's Shearwater, Yellow-crested Cockatoo; Endangered – Swift Parrot and Eastern Bristlebird; Vulnerable – Gould's Petrel, Cape Gannet, Superb Parrot, Cerulean Warbler, Marsh Grassbird; Near Threatened – Northern Bobwhite, Copper Pheasant, Oriental Darter, Black Rail, Bush Thick-knee, Plain Pigeon, Whistling Green-pigeon, New Zealand Pigeon, Red-headed Woodpecker, Olive-sided Flycatcher, Bell's Vireo, Flame Robin, Diamond Firetail, Golden-winged Warbler, Kirtland's Warbler, Brewer's Sparrow, and Painted Bunting.

In their 2020 continent-wide study across North America, Elmore and his colleagues writing in the journal *Conservation Biology* found the following species most vulnerable to collisions: Black-throated Blue Warbler, Ovenbird, Ruby-throated Hummingbird, Yellow-bellied Sapsucker, Wood Thrush, Brown Thrasher, White-breasted Nuthatch, American Goldfinch, Gray Catbird, and Common Yellowthroat. Their study also revealed migratory, insectivorous, and woodland inhabiting species identified as collision victims.

Conservationists have reminded all who will listen that the time to save a species is when it is abundant, not when it is on the brink of extinction or in trouble and threatened at any formally defined level of concern. Window strikes that kill individuals indiscriminately are an additive mortality factor that, to a mostly unknown degree, adversely affect and threaten the health of all bird life found within habitats containing human structures with sheet glass and plastic.

Chapter 4

Scientific Evidence – Making the Case

BEGINNING THAT WINTER DAY IN 1974 when I witnessed my first bird-window strike, I started to plan a comprehensive investigation about birds and windows. The first part of the plan was to determine what had already been documented in the popular and scientific literature. At the same time, I began systematically recording my observations and those of cooperating individuals. As my investigation matured, I added experiments to address the fundamental causes of bird-window collisions and how to prevent them.

All my experimenting was conducted outdoors. I used two general methods except for the first two preliminary experiments. One used a flight cage, now called tunnel testing by others who follow similar testing procedures but modified my original design. My experiments used sets of individual captured Dark-eyed Juncos, known to be frequent window-strike casualties. One by one, they were released at one end, and each directly flew to apparent safety, where habitat was visible through netting at the opposite end of the cage. The end where the subject flew to apparent safety and before the enclosing netting was divided into two equal halves. One half was unobstructed and served as a control. The other was altered by objects that the fleeing bird was expected to see and avoid by adjusting its flight path from the release point to pass through the unobstructed side.

The other method, a field test, consisted of framed windows that accurately simulate those in human dwellings. In this method, one clear or reflective window was unaltered and serves as a control. The others were clear or reflective and modified with objects expected to transform the space occupied by the window into a barrier that birds would see and avoid.

From the very beginning, the data I collected made it clear the killing was taking place at windows of commercial, residential, and institutional buildings in urban, suburban, and rural settings. Among my first questions were: who were the victims, where did they kill themselves, and under what circumstances? I expanded my searches on and around the campus of Southern Illinois University at Carbondale (SIU-C); in my immediate community, the small town of Carbondale with a population of about 30,000; and using information collected by a handful of select individuals at various locations who were passionately interested in birds and volunteered to do this on my behalf.

In my search of the scientific literature for bird-window collision accounts, I discovered one especially important article that allowed me to compile the very first comprehensive list of bird species killed striking glass. Fortuitously, in 1973, just prior to beginning my investigation, Richard C. Banks and his colleagues Mary H. Clench and Jon C. Barlow published a list of bird collections in the U.S. and Canada in *Auk*. This list was a treasure that gave me the opportunity to ask curators and collection managers about what window-killed specimens were cataloged in their museums. I also requested they share their expertise, informing me about their experiences and interpretations of any bird-window interactions.

I wrote to 466 curators and 11 individuals in the U.S. and Canada, asking for all their records about window-kills for the 1975–1976 period. I received 208 (44%) replies, and of these, 125 (26%) listed species, 13 (3%) estimated the number of annual kills but did not document species, 22 (5%) shared that they were aware of occurrences but offered no data, 8 (2%) stated they were unaware of any birds hitting windows, and 40 (8%) simply responded but sent no comment or data.

My survey of museums resulted in 125 replies, identifying 213 different species killed flying into windows. Of these, 207 were from North America north of Mexico. A member of my dissertation committee, Terence R. Anthony, recommended a technique to assess how complete my survey returns had been. He guided me to ask the question: What might we expect in terms of how many more different species would be documented as window-kills if I received more replies? The technique included dividing all the replies into five groups, each group consisting of a random selection of replies from five geographic regions covering Canada and the U.S. By identifying how many new species occurred in sequentially comparing the five groupings, this technique provided a means of estimating how many new species would be identified as

Chapter 4

window fatalities if the survey had 100 percent replies instead of the actual 44 percent. The analysis revealed that if all who received a survey had replied, an additional 10–20 species would have been documented, and the actual species list was about 89 percent complete.

Over the 42 years since this 1977 survey, I have obtained records of an additional 54 species. Considering taxonomic changes since the original survey, the 207 species (26.8 %) of the then estimated 775 North American species now includes 267 species (29.7 %) of the 898 regularly occurring North American species. That original survey was remarkably revealing and accurate. The species killed striking windows are those occurring around human dwellings. Those not on the dead list are those that do not or only rarely occur among human structures containing sheet glass. Unexpectedly, two species that were surprise casualties in the original survey or subsequently are the Band-rumped Storm-Petrel that hit a window in a stationary boat off Hawaii, and the Mallard that flew into a glass wall in the Netherlands and stimulated the formation of the Dead Duck Society.

The 14 window-killed species reported most often in the original and subsequent surveys, and those offered by volunteers, are American Robin, Dark-eyed Junco, Cedar Waxwing, Ovenbird, Swainson's Thrush, Northern Flicker, Hermit Thrush, Yellow-rumped Warbler, Northern Cardinal, Evening Grosbeak, White-throated Sparrow, Ruby-throated Hummingbird, Tennessee Warbler, and Yellow-bellied Sapsucker for North America; Blackcap, Garden Warbler, European Greenfinch, Chaffinch, Pied Flycatcher, Great Tit, Eurasian Blackbird, Eurasian Sparrowhawk, Song Thrush, Eurasian Bullfinch, Goldcrest, European Robin, Eurasian Woodcock, and Willow Warbler for Europe.

In southern Illinois, I made a special effort to gather as much detailed information as possible. Using university and regional newspapers, I solicited the general public, requesting all information anyone was willing to share. The window-kills my advisor Dr. George collected were cataloged in our departmental bird collection from 1971 to 1974. He also introduced me to several local residents who he was sure would be willing to gather data at their homes. Among these prospective collaborators were Jack and Muriel Hayward. Their home was lavishly covered with glass, and within the rural Union Hill community, located southwest of downtown Carbondale. Paul and

Carla Yambert's house was located some 8 km (5 mi) southeast of downtown Carbondale. The home of G. M. Brown was in a rural setting about 1.6 km (1 mi) southeast of the nearby village of Murphysboro. Each of these homeowners monitored and collected detailed records of window strikes at their respective residences. I also obtained bird strike accounts from 32 other local houses, 24 campus buildings, 4 commercial buildings and 1 unknown locality where the contributor did not record the strike site. All southern Illinois collection sites were within an area of 529 km^2 (204 mi^2) centered on downtown Carbondale in Jackson County, Illinois.

Hayward house in Carbondale, Illinois, showing façade with wall of windows, closeup of large windows, closeup of small windows, and view from inside looking out through large panes.

Among my records of southern Illinois window-kills is evidence supporting the species-specific vulnerability of the Yellow-billed Cuckoo that Townsend wrote about in his 1931 paper in *Auk*. From 1973 to 1976, I obtained records of 13 Yellow-billed Cuckoos striking windows of five different structures, four residences and one university building. Of these, 11 were fatal: four females, six males, and one unknown sex. All but one (September 18) hit from late May to August during the breeding season, suggesting that young that may have been depending on them for their survival were also lost.

My student and now colleague, Peter G. Saenger, is continuously in search of new species documented striking windows in the popular and scientific literature, from periodic solicitations, and from volunteers around the world that offer accounts to our Muhlenberg College Acopian Center for Ornithology website. As noted, our current world list consists of 1,348 (12.8 %) of the approximately 10,500 species of birds the world over.

Rothstein house in Purchase, New York, showing façade with walls of windows, closeup of large windows, and from outside looking through two clear panes on either side of an interior room.

Two homes, one in southern Illinois and the other in New York, were described in detail and continuously monitored for bird strikes. In both locations, the occupants primarily worked at home and were dedicated to recording bird strikes throughout the day and year-round. The southern Illinois home was that of Jack and Muriel Hayward, and they or their friends, Mr. and Mrs. Dwayne Dickerson, lived in this single-story ranch house and collected the data. Efrem Rosen, one of my graduate advisors at Hofstra University, introduced my studies to Poly Rothstein, who also lived in a lavishly glass-covered two-story home located about 1.6 km (1 mi) southeast of the town of Purchase, Westchester County, New York. The Hayward house was monitored from September 1974 to December 1976. The Rothstein house from August 1975 to December 1976. For each house, every window was numbered and the area, orientation, and facing habitat recorded. The Hayward house was one story, set on a slope and surrounded by shrubs, woods, and a lawn interspersed with trees. It has 52 windows ranging in size from 0.6–2.7 m (2–9 ft) wide by 0.9–2.1 m (3–7 ft) high, with a combined outer surface area of 114.5 m^2 (1,233 ft^2). The two-story Rothstein house had similar sized windows and surface area, set in a suburban neighborhood surrounded by open lawn, a few scattered large trees, and shrubs separating it from adjacent homes. The respective residents were given data forms upon which they recorded essential information for every documented strike. The information requested consisted of species, window number, date and time, habitat from which bird flew into the window, weather conditions such as temperature, sun condition, cloud cover, precipitation, wind speed and direction at time of strike. Additional information included notes on any noticeable disturbance near window strike, visible injuries, killed outright or seemingly injured, and if injured, behavior of the victim, amount of time required to recuperate and leave the impact site.

As I neared the mid-point in my graduate research, answering the fundamental question—Why do they do it?—became a necessary goal. It was now 1976 and I had collected a great deal of observational data. I believed I had a good understanding of the subject. The extensive evidence I possessed suggested the reason they did it was that they just do not see windows. With confidence, I met with my advisor and recommended that I had enough to write up and complete my dissertation. I recall being surprised by his reply, informing me that he expected an experimental component to be a prominent part of my doctoral research. My evidence to date was highly suggestive, but controlled experiments would be required to confirm my interpretations and understanding.

Chapter 4

    I immediately began thinking about what needed to be done, and I was aided by his thoughts. He explained that on October 28, 1976, he had a conversation with Dr. Willard D. Klimstra, a member of my dissertation committee. Dr. Klimstra asked: "Will birds fly into glass alone?" He wondered if some habitat or building features associated with windows, or a combination of such features, explained bird strikes. Dr. George said the same question was asked by his colleague and friend Wesley Lanyon, affiliated with the American Museum of Natural History in New York City, when they attended a Raptor Research meeting in November 1976. They concurred that setting out glass in "virgin" habitat without buildings around was the way to go. With these ideas as guidance, it was easy to craft my beginning experiments. Well, easy at least for the first one, but there were design and logistical challenges for others.

    My first preliminary experiment addressed what my advisors and his consulting colleagues wanted me to find out: Would birds fly into clear or reflective windows if not associated with a human building? Dr. George offered me six clear storm windows measuring 0.4 m (1.2 ft) wide by 1.2 m (4.0 ft) high, with a combined surface area of 2.9 m^2 (31.3 ft^2). Because of the spacing of vegetation, I separated each pane from 0–15.2 cm (0-6 in), wiring them against saplings on the periphery of a woody thicket facing an old field. When viewed from either side, except for a thin metal-framed border around each window, habitat was visible behind each window. No reflections were apparent off any of the panes. The windows were checked daily for eight days, November 8–15, 1976. During that period four birds were killed striking the glass: a Red-bellied Woodpecker, White-throated Sparrow, and two Northern Cardinals. I used this same setup with a reflective window. One of my volunteers, Mrs. Rene Potter of Makanda, Illinois, gave me a discarded mirror to simulate a reflective window. It was only one pane, measuring 0.6 m (1.8 ft) wide by 1.7 m (5.7 ft) high, with a surface area of 1.0 m^2 (10.8 ft^2). I placed it in the same location as the clear panes. Except for a few days when ice covered its surface during the early morning, a perfect reflection of field habitat and sky was visible on the pane when viewed from the field. The opaque, non-reflective side of the mirror faced the wood thicket. Daily checks over 12 days, November 19 to December 1, 1976, found two collision victims beneath the mirror, a White-throated Sparrow and Dark-eyed Junco. These first two experiments clearly documented that birds failed to see transparent and reflective windows that were not installed in human-built structures. Stated another way, windows need not be associated with man-made structures to kill birds. These results also further supported my interpretation that the avian visual system was not able to see clear and reflective windows as barriers to be avoided.

The materials in these first experiments were free, but it was clear that to investigate further, especially how to prevent bird-window collisions would require funding. My first thought was to build small mobile enclosures, like the models used to show what a sunroom addition to a home would look like. The plan was to create several of these units that could be moved and monitored at differing habitat sites. Although this was a terrific idea, the costs of this experimental design revealed it was not practical. I had limited personal funds, and the prospect of trying to attract outside grant funding felt overwhelming to me at the time. An alternative plan using cheap lumber, hardware to bind it, and salvaged glass was possible with the limited funds my wife and I possessed. With her agreement, I made the purchases and began constructing the flight cage and the simulated window units for the field experiments.

The total cost for construction was $695.64. This was a big burden on our finances in the spring of 1976. I felt obligated to reimburse the amount by seeking a research grant, but without experience and left on my own, it took a year for me to learn just enough to act. I was encouraged by a Pittsburgh Plate Glass Industries, Inc. (PPG) ad in a *National Geographic* magazine my in-laws had given me. The ad included two large sliding glass doors opening onto a patio, but with no dead birds beneath the panes, like those I had collected at many such sites. The PPG corporation seemed a natural source to ask for support. My experimental results would offer information on how hazardous their product was for birds, with the hope of encouraging them to make their sheet glass more environmentally friendly, specifically to protect birds. I used our library to investigate PPG and learned that they annually spent over $56 million on research. My initial and naive thought was that PPG could afford to reimburse me the approximately $700 for work that could inform them about the environmental impact of their glass on birds. I explained that I had already spent the money, had receipts documenting the purchases, and assured them that they were just supporting my cost of materials, no salary.

Chapter 4

In March 1977, I sent my inquiry to PPG's vice president for research, C. P. Blahous, at corporate headquarters in Pittsburgh. My unopened letter was returned with a note on the envelope stating he was not at this address. I immediately typed out another envelope and included next to the address "Please Forward To Where He Is" before mailing. In mid-April I received a reply in which Mr. Blahous politely explained that his responsibilities dealt strictly in creative glass construction and did not include my kind of request. He said he appreciated learning about my concern for birds and directed me to send my request to Grace K. Voegler, administrative officer with the PPG Industries Foundation. In her May reply to my letter she agreed to bring my request before the foundation screening committee after I provided a bit more information. She described how their foundation could not make grants to individuals. Their grants went only to institutions having a 501 (c) and 509 (a) (1) certification from the Internal Revenue Service (IRS). If I could supply the necessary information she would process my request. With great hope, I made an appointment with the SIU-C grants office to explain my need for these essential supporting documents to obtain my reimbursement.

Century-old barn on the farm of William G. George in Cobden, Illinois, before clear and tinted windows were installed.

My meeting with the SIU-C grants officer was not pleasant. After being offered a seat, I was sternly lectured on the proper procedures for applying for grants. It was supposed to start with them. I was informed about several required procedures and that I had violated all of them. Consequently, I was told they had no obligation to help me. I was crushed. Composing myself, I apologized for not knowing better, and asked if I was being told the university could not help me. The officer said no, they would help. What followed was a respectful collaboration with SIU-C Vice President for Academic Affairs and Research Frank E. Horton, Department of Zoology Chair William M. Lewis, my advisor Dr. George, and Grace Voegler at the PPG foundation. In July, SIU-C received a check in the amount I requested.

I received the requested reimbursement in October, minus $5.23 I was required to return in November, when the SIU-C accounting office informed me I was overdrawn on my grant. Undoubtedly, countless others have spent far more effort for far less, but through this first attempt to obtain research funding, I learned a great deal that would benefit my preparation of research grants thereafter.

By invitation, my second preliminary outdoor experiment was conducted by modifying a century-old barn on my advisor's farm. The barn had no windows. For the several years he occupied the property, Dr. George could attest, no birds flew into its opaque wood walls. I installed two clear and two gray tinted panes 1.4 m (4.5 ft) wide by 1.2 m (4 ft) high, having a combined surface area of 6.8 m^2 (72 ft^2) with their base 3.8 m (12. 5 ft) above ground. The clear panes were placed such that wooded habitat that came within a few meters of all four windows could be seen by looking through both clear panes from outside the barn. When viewed from an angle that did not permit a view through both panes, they appeared to reflect perfectly the facing wooded habitat. The tinted panes were positioned next to the clear panes. When viewed from any angle, they perfectly reflected the facing wooded habitat. Cloth trays were placed under each window to catch potential collision casualties. Each window was checked daily at various times of the day over one year, from February 1977 to February 1978. Once the windows were installed, birds hit the clear and tinted panes. Seven strikes were documented: at the clear windows a Bay-breasted Warbler and evidence of three unknown individuals, at the tinted windows an American Robin, Dark-eyed Junco and one unknown individual. This experiment provided more evidence that birds were not able to see clear and reflective panes, collisions occur wherever birds and windows coexist, and sheet glass rather than any other structural feature in which it is installed is the fatal hazard.

Chapter 4

Two separate designs comprised all of my other experimental protocols. As noted, one used wild-caught Dark-eyed Juncos in a series of decision trials in a flight cage, now referred to as tunnel testing. In December 1976, Dr. George asked Dr. Klimstra if he would permit me to use an area of campus assigned to the Wildlife Cooperative Research Laboratory to conduct my flight cage experiments. He agreed, and these experiments were done on a small patch of open land adjacent to a large fenced-in enclosure used to study rabbits, near the kennel and runs of our university mascot dogs, Salukis, a longhaired and sleek greyhound-like breed. The other protocol consisted of field experiments, and Dr. George invited me to conduct these on his farm in the Shawnee Hills of southern Illinois. The farm was 1.7 km (1.1 mi) north of Cobden, Union County, Illinois. Typical of small farms in this area, the land was covered with patches of wooded, field, and water habitats. Specific cover consisted of shrubs, conifer and deciduous stands, a small apple orchard, a sizeable lawn around the farmhouse, a cornfield, and two water impoundments.

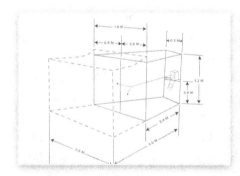

Flight cage used in experiments to test the ability of individual birds to see windows and objects placed on or near the glass surface to prevent strikes. Diagram depicting cage dimensions, actual cage under construction, and view from release box to end where test subject chooses to fly through an unobstructed (control) or obstructed (hawk silhouette) side to apparent safety.

The Dark-eyed Junco subjects used in the flight cage experiments were easily caught and were housed and cared for in small cages. Six individuals were used in each flight cage test that was run during daylight, from 7 a.m. to 6 p.m. My first test determined if birds, represented by juncos, could discriminate between clear glass and unobstructed air space. I then planned to use the flight cage to evaluate the effectiveness of various methods to prevent window strikes. Most of the prevention methods were also assessed in the field experiments. To keep the birds under a minimum amount of stress, any one subject was exposed to only one test method on any test day.

The flight cage was trapezoidal in shape. The floor was earthen and partly grass-covered. The sides for the first 2.4 m (8 ft) consisted of opaque brown masonite sheets. The outside of the rest of the enclosure consisted of aluminum screening stapled to a 2.5 by 5.1 cm (1 by 2 in) lumber frame. The roof above the masonite sides was covered with a white translucent cloth to illuminate the interior and match the outside lighting. Inside the narrow end of the cage was a small cardboard box measuring 14 by 8 cm (5.5 by 3.1 in), and located 0.6 m (2 ft) above the floor. At the other end of the cage, 2.3 m (7.6 ft) from the narrow end, the space was divided into equal halves measuring 0.9 m wide by 1.2 m high (2.9 by 4 ft). Each half was separated from the other by a 2.5 by 5.1 cm (1 by 2 in) wooden lath. During testing, subjects were placed in the box at the narrow end of the cage through a door to the outside. An opposite door in the box had a transparent window facing into the cage, which measured 4 by 8 cm (1.5—3.1 in) to stimulate the subject bird to face toward the broad end of the cage. A test trial began with opening the containment box door into the cage. The junco immediately flew to the opposite end, where it would have to choose which half to take as a route toward the wooded habitat visible outside the broad end. Each test consisted of at least ten trials. Additional trials were conducted when the ability of a subject to discriminate between the two halves was in doubt. An example was when a bird seemed to show a preference for only one side of the cage. The obstructed item and the unobstructed control were switched after five trials with each bird being tested, to address a suspected bias.

A single trial consisted of (1) placing a subject bird in the release box and allowing it to calm down, (2) opening the release box door into the cage, (3) observing which half of the cage the subject flew through, and (4) capturing

Chapter 4

the subject by hand and either placing it back into the release box for the next trial or placing it back into a holding bag if the test was completed. At the far end of the cage, birds were contained for recapture after a trial by nylon mist netting behind the obstructed and unobstructed halves.

Clear glass covered the obstructed side of the flight cage only in the first experiment. This initial test was to determine the ability of subjects to discriminate between clear sheet glass and unobstructed airspace. Subjects were not able to discriminate. In all subsequent experiments, clear glass was not used, to prevent subjects from being injured by striking the window surface. The unobstructed airspace simulated a clear window and served as the control. Adjacent to the control, single and multiple patterned items being evaluated to prevent collisions were either hung with clear monofilament line, attached to the wooden frame, or placed on the ground in front of the obstructed side.

First 1976 field experiment site and individual wooden-framed windows simulating those in houses along a tree line facing a cornfield, and current field experiment site.

All field experiments consisted of simulated wooden-framed picture windows as installed in human structures. In the first experiment, clear glass was installed in five units to determine the ability of birds to distinguish unobstructed habitat from habitat seen through clear windows. The window units were placed along a wooded hedgerow facing a cornfield, as if installed in new houses, and where no other man-made structure previously existed. The units were identical, and the dimensions of the windows were 1.4 by 1.2 m (4.5 by 4 ft) with their base 1.2 m (4 ft) above ground, having a combined surface area of 8.5 m^2 (90 ft^2). The distance between units ranged from 12.9 m (43.3 ft) to 23.8 m (78 ft). When viewed from either the cornfield or woods, habitat was visible behind the glass, or reflected on the glass surface. Depending on time of day, a combination of these effects occurred, depending on lighting conditions and the angle of view. Wire mesh trays were placed under each window to catch collision casualties.

The first field experiment was 20 days, from March 19 to April 7, 1977. All windows were checked daily at dusk for any evidence of a bird strike. Strikes were recorded when a specimen or specimen remnant, such as blood, other body fluids, feathers, or body and feather imprints, were seen on the glass surface. Thirteen strikes were recorded over the experimental period. Eight (61.5%) were fatal. Strikes were distributed evenly over the five units, indicating no bias in unit location. The outcome of this first experiment indicated that the same design could be used to evaluate various means of preventing bird-window collisions.

To address window location bias, for at least two decades, the control and test (treatment) panes in the field experiments were assigned randomly and physically carried to a new position at the end of each day. Moving these wooden-framed windows was hard work and caused stress, strains, even shoulder separations. My part in the experiments resulted in a few aches and pains, but it was my research assistant and colleague, Peter G. Saenger, who bore the brunt of the physical effort. I share his part in the experimenting because of his brilliant addition to our experimental design that helped us conduct each experiment more efficiently and, most importantly, preserve his health and, to a more minor degree, my own. His idea was to design and implement a sliding track system by which, each day, the study windows could be moved the way a sliding barn door is moved, instead of having to be carried from one daily position to the next. This improvement continues as an integral part of our field experiments, and no other part of our testing of bird-window collision deterrents has benefited our work and our bodies more.

My local search for window casualties and work to construct enclosures for flight cage subjects provided additional evidence that birds behave as if clear windows are invisible to them. Both can be considered additional evidence to support the results of the flight cage and field experimental protocols.

Emaciated remains of American Kestrel in an abandoned rural house in southern Illinois.

In more than one check of abandoned rural homes in southern Illinois, I found the emaciated bodies of American Kestrels in second-floor rooms. I speculate that these victims first entered the house from an opening in the roof, seeking a nest or roost site. Once inside, the light from an open trap door shining into the dark attic lured them down to the second floor. From there, they were attracted into a room with clear windows through which they could see familiar habitat. At each such discovery, the scene in these rooms was grisly and haunting when one thought about how the victims died. The windows and sills were soiled with smudges and waste. It was evident that the victims repeatedly flew against the window, attempting to reach habitat that was there before their very eyes. Their persistence in attempting to reach freedom indicates they were unable to retrace their movements to return safely outside. Their futile actions and eventual starvation suggest these birds could not comprehend the window's invisible barrier to freedom.

The other evidence derived from my experience with a wire mesh flight cage the size of a small room 6.1 m by 3 m (20 by 10 ft), which had an entrance the size of a conventional room door 0.8 m by 2.1 m (2.6 by 7 ft). I left the door open and sprinkled a seed trail from outside through the door and into the cage. Throughout the day, ground-feeding species, including White-throated and Song sparrows, Dark-eyed Juncos, and House Finches, foraged along the trail and entered the enclosure. Once inside, the birds repeatedly flew against the cage walls with force, as if trying to reach habitat seen through the wire mesh. The birds remained inside and continued to fly against the cage walls, even though the cage door remained open, providing an unobstructed passageway to safety. Only on very rare occasion, seemingly by chance and not guided by purposeful movement, did an individual fly through the open door after repeatedly flying against the mesh near the opening.

These birds behaved as if viewing and trying to reach habitat at a distance seen through the mesh. Like the house-trapped kestrels, they seemed unable to retrace their movements back to the entrance or understand why they were prevented from flying to habitat before them. Also like the kestrels, the lure and attention of distant habitat likely explains their inability to see the wire cloth in front of them.

The species-specific characteristics of victims are probably the best means of explaining why some species are more vulnerable to windows than others. A 2015 paper on bird-window collision victims during the spring of 2009 and 2010 in Toronto by Marine Cusa and her colleagues, writing in the journal *Urban Ecosystems*, was the first to detail certain traits that best explain why some species or species groups—what ecologists call *guilds*—are more likely to strike and be killed at windows. They found that individuals hitting windows in urban areas with rich surrounding vegetation were species of forested habitat and forest gleaners. Among others, the Blue-headed Vireo, Canada Warbler, and Golden-crowned Kinglet were frequent casualties in these settings. In contrast to these results, they found birds hitting windows in urban areas with less vegetation and large open areas of pavement and roads were species of open woodland spaces and ground foragers. Among others, Blue Jay, Ovenbird, and White-throated Sparrow were frequent casualties in these landscapes. The more we learn about what window-kills have in common, the more we will be able to identify why some species are more vulnerable. Investigations revealing species-specific traits of collision casualties are to be encouraged for the value these studies offer to avian conservation.

I am frequently asked what best explains who gets killed in what numbers at any one window collision location. My answer is consistently the same, and that is the density of individual birds in the immediate vicinity of the glass surface. In 2008, Steve Hager and his colleagues writing in the *Wilson Journal of Ornithology* stated that their test of a density effect was not supported by studying bird-window collisions on the Augustana College campus in Rock Island, Illinois. They assessed that the amount of glass, presence of vegetation, and behavior better explained the number of casualties they discovered. All of these factors certainly are relevant to explaining bird-window strikes. Their definition and measure of density, however, is problematic. To measure density, they used point counts within 50 meters of the windows they monitored. In my view, this is too great a distance. A more accurate and meaningful definition of density near an invisible hazard is the number of individual birds in the immediate vicinity of a window, the area within 10 meters (33 ft) of the glass surface. Within this distance, individuals are in a danger zone where they are more susceptible to being deceived by the glass surface. The results of a 1993 study by Erica Dunn writing in the *Journal of Field Ornithology* found more strikes occurred at windows near winter-feeding stations with greater numbers of birds. The attraction of feeding stations most often explains high densities of birds within this danger zone in front of residential and wildlife area visitor center windows.

My combined data from southern Illinois and records from those single homes in Purchase, New York, and Carbondale, Illinois, revealed that there is no greater risk of a window strike based on age or sex, for all or select species analyzed. A 2014 study by Steve Hager and Matthew Craig, writing in the online journal *PeerJ* in 2014, reported the vulnerability of summer breeding birds to windows in northwestern Illinois. They found juveniles were more vulnerable than adults for the most abundant species, American Robin; adults were more vulnerable than juveniles for the least abundant species, Red-eyed Vireo. More detailed studies are required to reveal what, if any, species-specific age and sex vulnerability occurs for collision casualties. From what we now know, adults and immatures, males and females are at least equally susceptible to striking windows.

Accounts in the ornithological literature and my records specifically reveal various ways behavior increases the vulnerability of different bird species to windows. Alternatively, for some few, behavior can protect a species. At least three species seem to be protected from striking windows even though they frequently occur in large numbers around human dwellings. They are Rock Pigeon, European Starling, and House Sparrow. All three have been killed flying into windows. However, the rarity with which these fatal strikes occur given their abundance near large amounts of sheet glass in urban settings suggests they are largely immune. Their flight habits around buildings likely protect them from injury. Their landing, take off, and general slow flight reduces the force with which they hit. Another means of protection is experience. Once a non-lethal strike has occurred that individual may learn that the space the window occupies is to be avoided. All three of these species seem to avoid windows by hovering in front of or slowly flying to nearby perches, such as the sills in front of windows or wall-covering ivy.

One account I obtained reinforces that no matter how infrequently these birds hit windows, they are still vulnerable and capable of being deceived under select circumstances. Over lunch at a professional meeting, Richard Johnston described to me how Rock Pigeons were frequently killed by flying against small, recessed, cave-like windows of the Museum of Natural History building on the campus of the University of Kansas. He explained how the pigeons wheeled around and would swiftly enter these cave-like entrances, where they were met and killed hitting the glass just inside. From the bird literature, in his classic 1940 life histories, Arthur Cleveland Bent described how hummingbirds learned to avoid the glass sides of cages. Similarly, individuals of various species living near buildings may benefit from familiarity of habitat that offers protection and learning from non-lethal strikes. In a 2016 study by Ann Sabo and her colleagues, writing in the online journal *PeerJ*, suggested such protection for resident birds in Norfolk, Virginia. They judged that the residents in their study site learned to avoid and thereby reduce their likelihood of striking windows. If this type of learning occurs in the wild, it may serve to protect at least some individuals, but it is likely to be of limited consequences for most.

For many species, however, their actions have dire consequences around windows. Those that fly regularly through restricted passageways in heavy cover are killed attempting to reach lighted areas behind or reflected in glass. Frequent among this group are *Accipiter* hawks, grouse, thrushes, and waxwings. Those species known to fly through open doors and uncovered windows have greater risk.

In 1962, Hans Lohrl writing in the journal *Kosmos* described how Barn Swallows were especially vulnerable to windows. He reported that window-killed individuals from one colony were found daily at a nearby clear glass corridor until the colony was reduced to such a point that it was abandoned. What makes this species particularly vulnerable is their habit of swiftly flying in and out of unobstructed doors, windows, and various types of overhangs. When glass covers these areas, they fail to see it as a barrier. They almost certainly have little or no chance to learn to avoid these invisible barriers by experience because they strike with enough force to cause instant death or sustain a fatal injury.

As with humans that walk into glass doors, distracted birds may be at additional risk to an unseen barrier. Distracted victims are individuals chasing one another, individuals escaping danger, predators pursuing prey, individuals intoxicated and impaired from eating fermented fruits, and those that are spatially disoriented due to a combination of adverse weather and artificial lighting.

Birds, like moths, can be drawn to light. The Toronto-based Fatal Light Awareness Program (FLAP) was originally named so because of the interpretation that nocturnal migrants were attracted to urban lighting and killed themselves flying into illuminated buildings. Lights Out programs in several North American cities have similar origins. Certainly, some individuals probably die at night by colliding with opaque or glass parts of lighted high-rises or other birds swirling around the tops of lit buildings. These fatalities, however, are rare events. Concentrated swarms of nocturnal migrants occur under adverse weather conditions, when cloud cover forces passage birds to fly at low altitudes. Flying at these lower levels, the lights of cities can reach and attract them. Under most weather conditions, however, and especially on clear nights, birds fly high enough not to be affected by city lighting. When they are attracted and congregate about lighted structures, migrants move in and out of lit areas, typically acting as if disoriented, becoming exhausted and fluttering to the ground. Once on the ground, birds are forced to move within the city's canyons of concrete and glass. They are now vulnerable to being deceived by clear and reflective panes. They kill themselves trying to reach the illusion of vegetation reflected in or seen behind windows in buildings, atria, through corridors (linkways) or glass panel railings along walkways.

Planted tree behind window in downtown Chicago, Illinois.

A 2006 study of spring migrants at the top of the Empire State Building in New York City by Robert DeCandido and Deborah Allen, writing in the journal *The Kingbird*, document the non-lethal hazard that lights pose to birds at this iconic skyscraper. Similar events have occurred elsewhere in North America, and careful study over decades has revealed to FLAP in Toronto that window strikes are almost exclusively daytime, not nighttime, events.

Annual display of dead by collisions in Toronto, Ontario, Canada, assembled by the Fatal Light Awareness Program (FLAP) to inform the public about the need to protect birds from windows.

Select environmental components explain how some windows are more hazardous to birds than others in residential and commercial buildings. These components include surrounding landscape, the size, height, and clear or reflective panes, location in urban, suburban and rural habitat, presence or absence of attractants such as vegetation, water and food, season, time of day, and weather.

Several published works have identified that the amount of glass surface and the presence of nearby vegetation best explain the number of window-kills at collision sites. Birds hit windows wherever they occur, but strike rates were highest in suburban and rural environments, which in most cases typically contain the greater bird densities near human structures. Looking specifically at the window threat to birds in urban areas, in 2013 Steve Hager and his colleagues, writing in the online journal *PLoS One*, reported bird strikes were correlated positively to window area and negatively to development. The more glass, the more bird kills. A greater number of buildings in a concentrated area without intervening green space results in fewer birds and fewer killed.

Increased casualties occur at buildings with nearby vegetation in the form of trees and shrubbery that attract birds to the danger zone near windows. Documentation and analyses describing how windows are made more fatal for birds if there is nearby vegetation are presented in the published investigations by Steve Hager and his co-authors writing in the journal *PLoS One* for northwestern Illinois in 2013, Yigal Gelb and Nicole Delacretaz writing in the journal *Northeastern Naturalist*, and my co-authors writing in the *Wilson Journal of Ornithology* in 2009 for New York City, and Cusa and her co-authors writing in the journal *Urban Ecosystems* for Toronto in 2015. Scott Loss and his colleagues, also writing in *PLoS One* in 2019, describe the principal collision predictors to be glass area and amount of surrounding vegetation in a study of downtown Minneapolis, Minnesota, which included the U.S. Bank Stadium, home of the Vikings NFL team.

Observations and controlled experiments support the interpretation that birds are equally vulnerable to clear or reflective panes. Most fatal collisions, however, occur at reflective panes, because most provide a deceptive illusion of the facing habitat and sky. Most windows are installed so that they cover dark interior spaces, while the surrounding area of the outer surface is more highly illuminated. Architects label the outer part of a window surface facing the environment as Surface #1. The brighter light outside interacts with the dimmer light to result in a perfectly clear pane reflecting the facing habitat and sky off Surface #1, as if it was a mirror.

My detailed studies at a variety of building heights reveal bird strikes were greater at large (> 2 m^2 = 21.5 ft^2) windows at ground level and at heights above 3 m (10 ft). In general, these are situations where vegetation, such as trees, is reflected in windows as high as four stories are lethal for birds, as they interpret the reflected illusion as a real tree. By contrast and at far fewer sites, birds are deceived and killed trying to fly to habitat on the other side of clear panes. As previously described, among other installations, these see-through effects occur at corridors (linkways), atria, outdoor railings, and noise barriers along highways.

Attractants that increase the density of birds near windows, and as a consequence increase the risk of window strikes, are feeders, fruiting trees and shrubs, water supplies in the form of bird baths and impoundments, nesting and perching sites in vegetation, and areas that offer protection from adverse weather conditions. The frequency with which finches, blackbirds, chickadees, titmice, nuthatches, woodpeckers, and hummingbirds are reported as victims is likely best explained by their regular abundance at feeders.

The frequency of strikes in different seasons is best explained by the seasonal abundance of birds in human-modified environments. I found seasonal strike rates to be highly variable at the continuously monitored single dwellings in southern Illinois and New York. Higher winter collision rates at single houses in these regions, compared to those reported elsewhere, are best explained by local site differences. The major one is the presence of feeding stations that attract large numbers of winter residents. Seedeater casualties predominate during the winter in both locations. They were Dark-eyed Junco, White-crowned and White-throated sparrows, and Northern Cardinal. Strikes in fall and spring consisted mainly of migrant warblers, thrushes, waxwings, and finches. Relatively few birds hit windows in summer, probably because their movements were restricted to breeding territories and adjacent areas. Southern Illinois summer casualties were Ruby-throated Hummingbirds, White-eyed Vireo, a Nashville Warbler as an early migrant, and Yellow-billed Cuckoo. As noted before, the cuckoos were known breeders,

Chapter 4

determined by eggs in the oviduct of females. Also described earlier, the death of these breeding cuckoos almost certainly resulted in the death of their dependent eggs and young. Further confirming the importance of bird numbers near windows, several investigations support the conclusion that those species that occur in the greatest numbers during any one season will comprise the greater number of casualties for a particular location. Still another example of the rule that, the more birds in the vicinity of an invisible hazard, the more victims.

Detailed analyses of when birds strike windows during the day supported my initial hypothesis that more strikes will occur in the morning than at any other time. This hypothesis was confirmed by the strikes recorded at the intensely monitored residences in southern Illinois and New York. At these houses, more strikes did occur in early morning than at any other time of the day. A researcher from Indiana informed me that he and colleagues found window strike rates to be four times greater from sunrise to 1 p.m. than at any other time of day. But Jean and Dick Graber, renowned Illinois ornithologists who monitored bird strike rates at their vacation home in southern Illinois, found something different. They found strikes occurred more often from 10 a.m. to noon and 1 p.m. to 2 p.m. than at any other time of day. As my data revealed, most feeder watchers will attest that the largest concentration of birds at feeding stations near their homes typically occurs in early morning. In the community where the Grabers lived, however, they found that roaming flocks of feeder birds visited different residential home feeding stations at different times. At their home, the large numbers of strikes and resulting casualties occurred when the feeding flock visited their station in late morning to early afternoon. Here again, the best predictor of strike rate was the density of birds in the immediate vicinity of the hazard.

Various weather conditions, including lighting, have been hypothesized to increase window strikes. As previously described, light can enhance the deceptive effects of glass by hampering visibility and attracting potential victims to the vicinity of man-made structures. In 1968, Brynjulf Valum, writing in the journal *Sterna*, reported that bird strikes in Norway occurred more often at reflective windows receiving direct sunlight. In an unpublished 1976 independent research paper, S. Post and S. Witzler and her colleagues from Earlham College in

Indiana offered an opposing interpretation. They speculated that birds hit windows less often when there is glare off the glass surface. They concluded that birds were able to detect and avoid widows with reflective glare but provide no quantitative data to support this claim. My records from the continuously monitored houses in southern Illinois and New York document strikes that occurred under both completely sunny and completely overcast skies. There were slightly more strikes in direct sunlight. Collective records from several sources confirm that strikes occur with relatively equal frequency during sunny and overcast conditions.

Precipitation appears to influence the frequency of collisions by decreasing visibility and enhancing the attraction of lights at night. In 1963, Claus Konig, writing in the journal *Deutsche Sektion des Internationalem Rates für Vogelschultz, Bericht,* stated that in Germany birds were more vulnerable to windows on misty days. Accounts of fog and snow increasing strike rates were reported in 1963 by F. Carpenter and H. B. Lovell in Kentucky, writing in the journal *Kentucky Warbler*, and in 1972 by G. A. Hall in Virginia, writing in the journal *American Birds*. Both accounts described migrants being attracted to and hitting lighted windows. In addition to what was described for Toronto earlier, the influence of lights in attracting migrants to windows was also noted in a 2015 study in an urban park in New York City by Kaitlyn Parkins and her co-researchers, writing in the journal *Northeastern Naturalist*. Strike records from my continuously monitored houses reveal that birds hit windows under conditions of no precipitation as well as during rain and snow. One exceptional incident occurred at a small window of a rural home in Makanda, Illinois, in 1976. The home had a small window on the north side and several large picture windows on the south side. The placement of these panes created the illusion of a passageway through the house. During a severe snowstorm, an estimated 50-plus Dark-eyed Juncos were recorded flying into the small window by the homeowner Rene Potter.

Overall, most strikes occur during favorable weather, presumably as a consequence of the clarity with which habitat is visible behind or reflected in glass. Assessing all strike accounts, under conditions of poor visibility, day or night, birds may experience spatial disorientation, being especially vulnerable when attracted to lighted areas behind windows.

Data from the continuously monitored houses in southern Illinois and New York provided a means of determining whether window orientation affected strike frequency. I hypothesized that southbound migrants in fall would be more apt to hit north-facing windows. Northbound migrants in spring would be more likely to hit south-facing windows. This prediction was not supported at either location. Most casualties were nocturnal migrants killed during the daytime. During the day, when they were feeding and resting, their movements were not confined to a specific compass direction. Strikes occurred with relative equal frequency around the entire house. These results indicate that windows pose a lethal hazard for birds no matter what their orientation, and further support the conclusion that strikes are likely wherever birds and windows coexist.

Except for those two continuously monitored houses over a one-and-a-half-year period, most bird-window collision studies have been conducted at commercial or university campus buildings. A growing number of studies, however, are looking at single residential homes. Erin Bayne, his colleagues and students at the University of Alberta, have investigated window strikes at single dwellings. In 2012, writing in the journal *Wildlife Research*, they found the highest number of bird strikes and resulting mortality occurred at single houses with the following characteristics. In order of highest to lowest frequency, collisions occurred at rural homes with bird feeders, rural homes without feeders, urban homes with feeders, urban homes without feeders, and apartments. They recorded more fatal strikes in older neighborhoods. The greater number of fatalities in older neighborhoods was attributed to more mature and diverse vegetation that attracted more victims than the less-developed new housing areas. Additionally, highlighting the universal lethal threat of sheet glass, they documented window-kills at newly constructed houses with meager vegetation.

In 2016, Annie Bracey and her colleagues, writing in the *Wilson Journal of Ornithology*, recorded migrant window strike mortality at residential homes in Duluth, Minnesota, along the shores of Lake Superior from 2006 to 2009. They found that both fatalities and scavenging rates of victims increased with distance from the city center, estimating 11–16 bird kills per house. They found that the number of birds killed varied among the five migratory seasons they monitored along Minnesota

Point. More migrants were killed than residents. Small birds were likely to be removed more quickly than larger ones by scavengers.

I end this chapter with an account of a series of window strike fatalities at a sunroom addition to a home originally located in a rural setting but transitioning into a suburban environment. Since 1979, I have lived in the Lehigh Valley of Pennsylvania, about an hour's drive northwest of Philadelphia. Lehigh and Northampton counties are located in southeastern Pennsylvania and make up most of the Lehigh Valley. One fall day in 1983, Janet Bauer, a Northampton County resident in the village of Bath, called and asked if I would visit her for the purpose of seeing where a succession of window strikes occurred annually. Her records documented the strikes only during the same week in September, and included only the deaths of Black-capped Chickadees. Now, Black-capped Chickadee is almost always described as a permanent resident where they occur. Most birders do not think of them as migrants. Mrs. Bauer's records offered evidence to the contrary. She explained that she had remodeled her home by adding a glassed-in sunroom. Over several years after the sunroom was completed, during this one week in September, Black-capped Chickadees were killed in decreasing numbers from one year to the next. The first year, the sunroom's sheet glass walls claimed about a dozen chickadees. The next year, about ten fatal strikes occurred. In subsequent years, there were five, then three, then one and then none. These birds likely represented a migratory sub-population that regularly used the same flyway in their passage. I shared with her an account I had heard at a meeting by Peter Berthold, a renowned German ornithologist, who described the passageway fidelity of migrants. At his European study site, he caught the same Blackcap, an old-world warbler, in not only the same net but also the same level of the net one year after the next, until it was presumed it was no longer alive to migrate and be caught. I speculated with Mrs. Bauer how, perhaps similarly, her collected dead chickadees were also showing passageway fidelity on their migration until there were no more.

Chapter 5

What Happens When Birds Strike Windows – Cause of Death, Injuries, Recuperation

THERE ARE VERY FEW ACCOUNTS in the scientific literature of the injuries sustained by birds that flew into windows. In 1990 and 2005, writing in the *Journal of Field Ornithology*, my students and I documented collision injuries and assessed the cause of death. We recorded all external and internal (x-rays included) injuries of about 800 specimens. Birds that hit windows were judged to almost always hit headfirst. The resulting head trauma is the most reasonable cause of death.

Like humans, the brain tissue of birds is near the consistency of a banana. The brain similarly is confined to a space surrounded by bone within the skull. When the head strikes the unyielding glass surface, the force compresses the brain against the unyielding wall of the skull. The brain tissue responds by swelling. The confined swelling brain has nowhere to expand, except where the spinal cord enters at the base of the skull. The mounting pressure from the swelling results in breaking the blood brain barrier, and blood flows and pools within the brain tissue.

The result is unconsciousness, paralysis, or, for many, outright death. Skeletal fractures (broken bones) were present but relatively rare in all the casualties we examined. None of them sustained a cervical fracture (broken neck), but almost anyone picking up a fresh window-kill believes the victim died of a broken neck. They get this impression because of the flexibility of the avian neck. Birds have special saddle-shaped articulating surfaces as projections from their neck vertebrae that facilitate movement. Consequently, the limp neck muscles of fresh window-kills give the observer the impression that the victim has a broken neck because the neck seems abnormally flexible in every direction. If the victim is discovered hours after the strike, a broken neck is judged less likely because the neck is not as pliable because the muscles stiffen up due to *rigor mortis*.

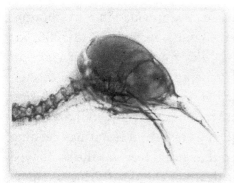

X-ray of window-killed Ovenbird showing upper and lower broken mandibles.

Four White-crowned Sparrows showing various amounts of intra-cranial hemorrhaging. All four birds died immediately after striking a window.

In a 2005 paper, Carl Veltri and I reported 82% and 91% of the kills resulting from collisions with a concrete nuclear plant cooling tower and windows, respectively, had no skeletal fractures. No neck vertebrae were discovered with any fractures in those victims that sustained skull fractures. My records from elsewhere, however, do contain one reliable account of a window strike resulting in a broken neck. The record is a pathology report from Tufts University Veterinary Medical Center shared by Mark Pokras. It contains a description of how an immature Peregrine Falcon was seen stooping on prey atop a high-rise building in downtown Boston, Massachusetts. The roof was a green space surrounded by clear glass walls. The falcon struck the inside transparent wall when it attempted to fly upward. Among other injuries, this high-speed collision resulted in the C6 vertebrae being crushed. The bird was initially paralyzed from the neck down, and succumbed from many strike-related complications of the impact.

Injuries to other body parts are also known to result from window collisions. The late Len Soucy oversaw the rehabilitation program for decades at the Raptor Trust in Millington, New Jersey. He described and gave me x-ray records of broken and displaced coracoids and pectoral girdle bones of window-killed and injured birds brought to his facility over the years.

Kit Chubb and her architect husband, Robin, built and established the Avian Care and Research Foundation in Verona, Ontario, about 32 km

(20 mi) northwest of Kingston. The care they provided in rehabilitating birds at their facility took place over 28 years, from 1978 to 2006. Kit Chubb's 1991 book *The Avian Ark: Tales from a Wild-bird Hospital* documents their experiences. She also prepared a detailed, unpublished special report in which she described her study of 397 window-collision casualties. She recorded the same head trauma injuries we found, but also internal hemorrhaging elsewhere, such as tears and resulting bleeding from the descending aorta. She noted that aortal tear was the same type of injury thought to have caused the death of the U.K.'s Princess Diana from a vehicle collision in France. She listed the following injuries from the collision cases she examined: 186 sustained brain damage, 118 had internal hemorrhaging, 48 had broken or dislocated coracoids, and 22 had evidence of cracked neck or other vertebrae. She also found broken furcula (clavicles), humerus, sternum, tibia, ribs, radius and ulna, tarsus, and femur. Four specimens had torn crops that she judged were full and burst like a balloon upon impact. Mrs. Chubb recorded ruptured and leaking air sacs resulting in puffiness under the skin, and interpreted these injuries resulting from a violent blow or puncture. Her patients were as likely to be adults (53%) as sub-adults (47%). All total, the collision victims she recorded as dead on arrival or treated consisted of 80 species. Thirteen of these were birds of prey: Sharp-shinned and Cooper's hawks, and Northern Goshawks. Other frequent window strike patients in her care were Ruffed Grouse, Mourning Dove, Downy and Hairy woodpeckers, White-breasted Nuthatch, and Rose-breasted Grosbeak.

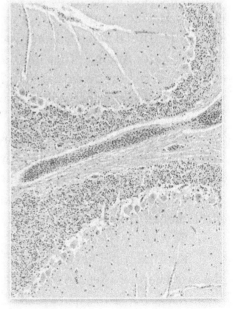

Histological section of brain of window-killed American Robin showing pool of blood within cerebellum.

Gathering additional evidence of head trauma injury, our research team prepared histological sections of the brain of two window-killed specimens, Sharp-shinned Hawk and American Robin. Both victims had extensive hemorrhaging in the cerebrum and cerebellum. Blood pools were most prominent in the cerebellar white matter. Fatalities resulting from collisions are most likely the result of damage to the cerebellar communicating fibers that are vital afferent and efferent tracts. As noted earlier, the rupturing of blood vessels within the brain at several sites is likely due to swelling and subsequent herniation of parts of the cerebellum and medulla through the foramen magnum. These and other specimens exhibited extensive subdural bleeding which caused and resulted in additional intracranial edema.

The brain injuries sustained by casualties qualify as fatal or near-fatal concussions. Symptoms exhibited by collision survivors support this conclusion. Prior to death, victims are often completely or intermittently non-responsive, and lack balance, normal posture or coordinated muscle action. Some exhibit ipsilateral drooping eye or wing, and dilated pupil. They frequently have rapid or slow heaving respiratory movements. To repeat for emphasis, these internal brain injuries best explain the cause of death of collision fatalities. Those treating survivors have had some success in administrating the drug dexamethasone sodium phosphate as much as six to eight hours after impact to help limit brain swelling.

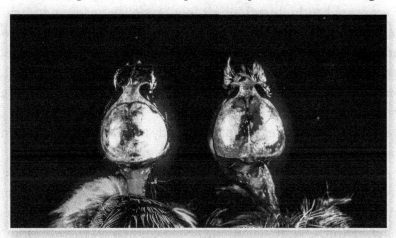

Yellow-breasted Chat and White-throated Sparrow that behaviorally appeared to survive and completely recover from striking a window exhibits massive intra-cranial hemorrhaging.

The consequences of window strikes vary greatly for individual birds. Some are killed instantly. Others are knocked unconscious or stunned and later succumb to their injuries or scavengers. Some recover enough to fly off weakly, but seemingly unharmed. Still others appear startled but unaffected and fly off immediately after striking. The most obvious external injuries are broken bills. The most obvious evidence of internal injuries is blood and fluid leaking from the mouth and nostrils. Internal examinations revealed that every strike fatality sustained some intracranial hemorrhaging.

The internal injuries sustained by birds that survived window strikes were determined by studying two specimens, Yellow-breasted Chat and White-throated Sparrow. Both birds appearing alert, active, and completely recovered two and four hours, respectively, after their collision. They were sacrificed to determine what if any injuries they appeared to overcome. Both sustained severe intracranial hemorrhaging, and the expectation is that both would have almost certainly died if released.

The recovery time of strike survivors ranged from a few seconds to days. The behavior of individuals that initially survived the effects of impact was generally similar. If knocked unconscious, they remained motionless, usually lying on their side and seemingly breathing slowly and regularly. After regaining consciousness or if only stunned initially, they sat upright with their body resting on their legs and feet. They appeared to be breathing deeply by slowly opening and closing their mouth. At this point, some either stopped moving their mouth, breathing appeared faint, eyes slowly closed, and they slumped and fell over dead. For others, heavy breathing and mouth movements became less frequent and exaggerated and eventually stopped. They appeared to be breathing without stress. After a period that varied with each bird, they raised up on their legs, and flew weakly but directly to the ground or a nearby perch within cover.

A few extraordinary accounts highlight the recovery power of injured birds. One of my students in southern Illinois worked at a hospital. He noticed upon reporting for work at 8 a.m. that an obvious window collision victim, a Virginia Rail, was lying below a plate glass window. It was cool outside. He made a mental note to collect the bird for me at the end of his eight-hour work day, which he did. He placed the carcass on the passenger front seat of his pickup truck and departed for home. Fifteen minutes into his drive, the truck cab warmed and apparently so did the bird. The revived rail flew about the truck until he pulled off the road and gained control of the panicked animal. He brought the rail to me the next day, and I kept it in captivity for observation for 12 hours. It recovered to appear alert, normally active, and flew off seemingly unhindered when I released it at our campus lake. This bird was unconscious for at least eight hours, survived, and recovered to appear healthy when returned back to the wild.

An account of an injured migrant offers evidence that birds can live to lead seemingly normal lives after experiencing a debilitating window strike. Michael Butler, a Canadian ornithology student keenly interested in the bird-window strike subject, documented and shared the account with me. I met Mike in 1983, during the 100th anniversary meeting of the American Ornithologists' Union at the American Museum of Natural History in New York City. He told me of an adult Indigo Bunting that hit a house window in Halton Region, Ontario, on May 13, 1975. The bird was knocked unconscious and stayed that way for a half-hour. The homeowner happened to be a bird-bander, and when the bunting appeared to recover completely he banded and released it.

One year later, almost to the day, the same bird hit the same window and killed itself, on May 15, 1976. This is a remarkable account if you consider that this individual likely went on to breed near or north of its accident and then traveled to its non-breeding grounds in Central or South America. Upon successfully returning to breed one year later, it was killed outright at the site of its original mishap.

In 2018, while attending the annual meeting of the Wilson Ornithological Society at the University of Tennessee at Chattanooga, David Aborn told me about a female Hooded Warbler window-kill he collected on that campus. It was a healthy wild-caught bird that had been banded two years earlier on its non-breeding ground in Belize. This female lived at least two and half years, making its neotropical migration at least twice before succumbing to a window collision. Certainly, these individuals provide irrefutable evidence that some individual birds can recover and survive a serious window strike, and yet remain vulnerable to the fatal consequences of what was for them an invisible hazard after surviving breeding and traveling thousands of miles migrating.

By contrast, my treatment of an Evening Grosbeak revealed the after-effects of collisions. This bird hit a window in southern Illinois and was brought to me by a concerned homeowner. She judged that its drooping wing indicated it had a broken bone. Like the rail before, I placed it in a cage for observation. Over the next three weeks it developed increasing paralysis. One morning when I arrived to observe and feed my charge, I found it lying on its side and rapidly moving its feet back and forth. It seemed to be running with nothing but air for traction. It could not care for itself in this condition. I sacrificed and examined it, conducting a necropsy. Necropsy is an autopsy on animals other than humans. I found a massive blood clot in the brain. I assessed that it was caused by the window strike that resulted in its progressive paralysis. That this victim sustained life-threatening neural damage from the brain clot was evident from x-rays revealing that there were no broken bones in either the drooping or non-drooping wings.

A curious and as yet unresolved consequence of window strikes for birds is feather loss in the head region. Andy Jones, Director of Science, William A. and Nancy R. Klamm Endowed Chair, and Head of the Department of Ornithology at the Cleveland Museum of Natural History, Ohio described the injury to me in 2012. While preparing specimens, he regularly noticed he could identify window-kills because they tend to have a line of feathers missing across the base of the throat. He speculated that upon hitting they either hyperextended their bills and neck upward causing a loss of feathers, or their heads dropped downward and the tip of the bill caused the feather loss. He favored the former explanation, and without capturing a high-speed video of a strike the cause of the missing throat feathers is still unresolved.

From those single houses my collaborators continuously monitored in southern Illinois and New York, they recorded that one in two (50%) individuals hitting their windows were killed outright. Other than the few accounts I shared here about those that eventually died or survived, there is no information on the rate or probability of recovery for those birds that hit and fly off after a collision. Almost certainly, most survivors are thought to perish, given the types of injuries strike casualties sustain. It is suspected that those that appear to recover completely also succumb later or are taken by predators or scavengers who target them due to their weakened condition.

Chapter 6

Valuing the Dead

Unless bird-safe glass is installed, the invisible lethal threat windows pose to free-flying wild birds will continue to occur, with more buildings being constructed in human-altered environments. The predictable and expected losses are a valuable source of information. Window-kills are a plentiful source of museum specimens, given that every piece of sheet glass and plastic in any human structure the year round and the world over is a potential killing site. Systematic searches of most human dwellings will result in the discovery of collision specimens in rural, suburban, and urban areas.

Window-killed Rose-Breasted Grosbeak and Nashville Warbler below a patio glass door.

Practical uses of window casualties include informing us about migratory movements and distribution, breeding and non-breeding ranges and their contractions and expansions, new occurrence records for geographic locations, and as subjects for whole or in part specimen-related studies of species-specific form and function. These and other sources of specimens are an important resource for art.

Writing about an endangered species in *American Birds* magazine, Walkinshaw reported a Kirtland's Warbler window-kill along its migratory route between Michigan and the Bahamas. If this discovery occurred before the long-sought knowledge of non-breeding grounds of this species, this specimen could have provided an essential clue. Using a straight edge from their nesting sites in northern Michigan through the kill site in Ohio to their overwintering area in the Bahamas would have narrowed the search.

In 1991, writing in the journal *Corella*, A. Talpin described how he used window-killed specimens to study the migration of many Australian birds. In 2005, in *North American Birds*, H. L. Jones reported the first record of a White-bellied Emerald, a hummingbird, in El Salvador, which was a window-collision victim in downtown San Salvador on November 3, 2004. In the *Journal of the Pennsylvania Academy of Science*, in 1982 and 1983, and continuing to the present, my students and I have used window-killed specimens to describe the gross anatomy and histology of the digestive tracts of American Robin, House Sparrow, and several other species.

Window-kills should be properly documented and preserved in authorized collections if we are unwilling or unable to stop the killing at clear and reflective panes. About 1,000 window-kills are annually added to the bird collections of the Field Museum in Chicago and the University of Nebraska State Museum in Lincoln. When I was first invited to Chicago in 2005, my tour around downtown dramatically revealed how window-killed specimens could enhance our knowledge about the pathways of migrants. I was amazed by the abundance of Brown Creeper and Lincoln's Sparrow that were among the dead and injured below and on sills in front of windows. These species occur in eastern Pennsylvania where I live, but nowhere near the number or concentration that I saw in Chicago, especially Lincoln's Sparrow. What other species-specific migratory movements can be revealed by paying attention to the ubiquitous source of specimens below the windows of the world?

Chapter 7

Why do We Care – the Services Birds Provide

IN 2013, EXPRESSING CONCERN AND emphasizing the interconnectedness of life, Travis Longcore and P. A. Smith, writing in the journal *Avian Conservation Ecology*, warned that avian deaths from windows and other human-associated mortality factors pose a growing deleterious effect on the world's ecosystems and the goods and services birds provide them.

In a money-centered world, those goods and services are getting increased attention, especially considering what it might cost to have humans try to do the same work. Pest control is one service birds provide that contributes to productive yields of valuable crops such as coffee and grapes. They also provide public health benefits such as consuming insect disease vectors and scavenging the dead. They play a role in pollination and seed dispersal and provide high-quality fertilizer from seabird guano. They also serve as ultra-sensitive indicators of environmental health on the local, regional, and global levels. The "web of life" that Alexander von Humboldt described as essential to the health and very existence of humankind is as relevant today as it was when he wrote about it two centuries ago. Like every other living being, birds are an essential part of that complex, interacting super-organism that encompasses all life.

One of the most dramatic pest-control events occurred in 1848, in what is now the city of Great Salt Lake, Utah. To many, the event was a true miracle. California Gulls (*Larus californicus*) descended on the so-called "Mormon crickets" (*Anabrus simplex*), an insect that is not actually a cricket but a katydid that grows to 8 cm (3 in) and voraciously consumes vegetation. Crops and even their own are on the menu during their swarming phase that can move across 2 km (1.2 mi) of agricultural fields in a day. These gulls are credited with saving about 4,000 Mormon pioneers by eating the katydids and preventing them from consuming their second harvest. As a grateful tribute, a monument to the California

Gull today stands prominently in Salt Lake City, commemorating the life-saving service of this bird to people.

The services of mosquito-eating birds contribute to limiting the spread and prevention of malaria, yellow fever, and other diseases. This is a particularly important aid in tropical climates, where several species, but especially martins and swallows, have protected indigenous people from the earliest times until now. Vultures the world over are specifically adapted to removing the dead, and with them the accompanying organisms that spread diseases among humans and other animals. At the turn of the 21st century, the dramatic disappearance of vultures in India virtually eliminated the service they provided in scavenging livestock carcasses. The consequence of this loss is an estimated 48,000 human deaths from rabies and a cost of $34 billion to the national economy. Other vulture-connected health-related costs of $24 billion were linked to the increase in scavenging feral dogs and rats that carry rabies and bubonic plague, respectively, in addition to other human-susceptible diseases.

Hummingbirds offer pollinating services for commercial flowering plants and many human foods. The dispersal of seeds by fruit-eating birds ensures reforestation, and with it ecological succession that consists of a chain of changes in habitat that provides homes to variously adapted life—including diverse bird species and the food and shelter they require to survive and sustain healthy populations.

One practical service birds provide is preying on insects across the boreal and temperate forests of North America. Given the number of species preying on insects, collectively the presence of birds can have meaningful consequences for the health of the trees in these forests. Martin Nyffleer and his colleagues, writing in the journal *The Science of Nature* in 2018, estimated that the world's insectivorous birds annually consume 400 to 500 million metric tons of insects per year. Forest birds account for 70% of this amount, or greater than 300 million tons a year. Especially for forests, the ecological and economic importance of birds eating harmful insect pests has tangible worldwide value.

With currently estimated populations numbers for Blackpoll and Bay-breasted Warblers at 60 and 9.2 million in the boreal and temperate North American forests, respectively, these two species are specific examples of valued insectivorous birds. Both are frequent window-kills, and based on 2008 estimates, over the past prior four decades the Bay-breasted Warbler has declined 64%, the Blackpoll Warbler by 65%. Although neither of these species is considered rare, their dramatic declining numbers indicate our forests are less healthy due to the presence of greater numbers of insect pests and the harm they exact. The annual monetary value of services provided by birds to boreal and temperate forests—for consuming insect pests and dispersing seeds—is estimated at $1.385 trillion worldwide, $90 billion for the U.S. alone.

Many trace the very beginnings of the U.S. environmental movement to birds, who have shown their value as a means of measuring environmental health. Credit is due especially to Rachael Carson and her 1962 book *Silent Spring*, which described the lethal effects of DDT and other poisons on birds, and by implication on us. Disappearing raptors and pelicans brought the harmful environmental effects of pesticides to the attention of conservationists. In 1972, scientists and activists finally succeeded in persuading the U.S. government to ban DDT.

For many, birds are prominent representatives of nature and even of human culture, thanks to their beauty, their association with good feelings and health, and the ways they often serve as symbols of cherished beliefs. There is no way to assign monetary value to such contributions to human life, which are literally priceless. In his 1977 book, *A Bird Watcher's Adventures in Tropical America*, the iconic botanist and ornithologist Alexander F. Skutch included birds among the elements that most delight human life on this planet. He wrote, "For a large and growing number of people, birds are the strongest bond with the living world of nature. They charm us with lovely plumage and melodious songs; our quest of them takes us to the fairest places; to find them and uncover some of their well-guarded secrets, we exert ourselves greatly and live intensely. In the measure that we appreciate and understand them and are grateful for our coexistence with them, we help to bring to fruition the age long travail that made them and us. This, I am convinced, is the highest significance of our relationship with birds."

The U.S. Fish and Wildlife Service (USFWS) periodically prepares a report that in part documents the economic value of birds, based on what can be measured. This report describes which interests U S. citizens spend most of their recreational money on. Gardening ranks first, but birding is second.

The latest accounting on the value of birding to the U.S. economy was in 2013, reporting that nearly 47 million birders spent a total of nearly $41 billion. This resulted in 666,000 jobs, producing more than $31 billion in employment income, and generating $6 billion , in state and $7 billion in federal tax revenue. Traveling and tourism, a major part of which is focused on wildlife watching and birding in particular, generated $7.6 trillion, or 10%, of the world's Gross Domestic Production (GDP). These recreational activities accounted for 277 million jobs, 1 in every 11 in the global economy. The interest in birds is big money. The sale of birdseed alone is a multi-billion-dollar industry in North America. Birds have earned our protection for both philosophical and financial reasons.

No one should have to live in a world without birds. No harmful products or actions should deprive our children of the opportunity to receive the services and joy birds provide.

Chapter 8

Getting the Word Out – Transforming Education into Action

From the 1970s to the present, scientific research papers and accompanying popular articles about bird-window collisions have continued to grow. Current publications document extensive details that include: (1) quantitative studies revealing species, building, and environmental conditions, (2) individual injuries and causes of death among strike victims, (3) the level and composition of mortality as a species-specific conservation concern, (4) means to prevent bird-window collisions, and (5) a number of reviews updating the latest body of collective knowledge. This growing body of literature has resulted in increased awareness and action among conservation-minded constituencies that now include saving birds from windows as part of their mission. Just as important is the awareness among building industry professionals. A growing number of educational programs targeting the building industry have increased awareness in recent years, but not enough to result in making human structures safe for birds as standard practice or even a regular design consideration.

Scholarly work can build a case for changing human behavior that in turn can be the basis for social change for the better. On this issue in particular, efforts have included trying to attract all relevant constituencies to join in a common cause of convincing people they have the power to make the built environment safe for other life. Specifically, to make sheet glass and plastic in all human structures safe for birds. Although much yet needs to be done, the process of informing and acting to make windows safe for birds has been rewarded with some success. The collective efforts of those dedicated to saving birds from windows surely qualify as an example of persistence as defined by Calvin Coolidge.

Chapter 8

> *"Nothing in the world can take the place of persistence. Talent will not; nothing is more common than unsuccessful men with talent. Genius will not; unrewarded genius is almost a proverb. Education will not; the world is full of educated derelicts. Persistence and determination alone are omnipotent. The slogan 'Press On' has solved and always will solve the problems of the human race."*

From my orientation, background, and bias, scholarly work must come first to document and establish a need for change. At the same time, or ideally shortly thereafter, the science will convince and stimulate popular writings to reach a broad audience, one that eventually will enlist the general public to demand reasonable, justifiable changes that will benefit human society, which is part of and interdependent with nature.

For my part, I sought to publish my dissertation results in a peer-reviewed scientific journal immediately upon finishing my graduate degree in 1979. As it turned out, doing that required unexpected struggle, which took every bit of the persistence Coolidge described.

Naively, I believed that the most prestigious and wide-read general science and ornithological journals in the U.S. would be eager to publish my work, based on the novelty and importance of my findings. With enthusiastic hope and expectations, I submitted my first manuscript to *Science* and then *Auk*. The editors of both promptly returned my manuscript, telling me the topic was unsuitable for their respective publications. It took me until 1989 to finally convince the Wilson Ornithological Society that the topic was suitable, and with favorable peer reviews, their editor accepted my first overview paper describing the subject. It was printed in their journal *The Wilson Bulletin*, since renamed *The Wilson Journal of Ornithology*.

The detailed findings in my dissertation on the injuries birds sustained from window strikes, the level of attrition, and the means to prevent it still needed to be published. Hopeful but expecting I had to convince another editor that the topic was suitable, in early 1990 I submitted another manuscript to *Condor*, at the time arguably the second most prestigious ornithological journal in the U.S. after *Auk*. The editor sent my manuscript out for peer review but rejected it, even though it received two positive referee evaluations recommending publication.

As common as manuscript rejections are in science, especially for young researchers, I could have accepted being rejected for what seemed a reasonable cause, but yet again I was told the topic was unsuitable, with no clarifying explanation. This rejection was frustrating because it seemed inconsistent with my training. After all, I was taught and had believed that all topics were suitable for scientific investigation. To be sure, the study data, analysis, interpretation, and presentation had to be scientifically worthy. But to be rejected because the subject matter was judged inappropriate seemed unreasonable and wrong.

Elizabeth Kolbert explained the difficulty in changing your mind or being open-minded about what seemed unreasonable according to convention in her 2014 Pulitzer Prize–winning book *The Sixth Extinction*. She described how extinctions by catastrophism were considered unreasonable compared to the more accepted uniformitarism advocated by scientific luminaries such as Charles Darwin and Charles Lyell. Perhaps the editors who judged the window kill topic to be scientifically unsuitable did so because their experiences did not permit them to accept what were unbelievable results and interpreted claims. If so, it would have taken little effort to add that to my rejection letter. My interpretation of their conventional thinking was that the topic seemed unsuitable because they had no experience with it over their long and distinguished careers. I naively thought that if there was measureable evidence—and I objectively presented a great deal—there were no subjects prohibited to scientific inquiry.

The opportunity to publish occurred at the 1990 annual meeting of the Wilson Ornithological Society, where I met Edward (Jed) H. Burtt, Jr. We chatted about what each of us was studying, and I could not help but reveal my dismay, frustration, and discouragement trying to publish the rest of my dissertation results. Jed told me he was the current editor of the *Journal of Field Ornithology* and that he found the bird-window topic of high interest and value, based on the kills he discovered on his campus at Ohio Wesleyan University. He asked who the *Condor* referees were and said he knew both and trusted their judgment. He promised to evaluate the content from *Condor*'s review and conduct his own. The result was that he accepted the study, recommending that I divide the content into two papers, covering injuries in one and toll and prevention in the other. Both papers were printed in the 1990 volume of the *Journal of Field Ornithology*.

From these initial fundamental articles, attention to window-kills in the scientific literature has grown. From 1994 to 1998, I corresponded with Timothy J. O'Connell and supported his efforts to publish a bird-window kill study in Virginia. Tim monitored bird strikes at an apartment complex in Richmond while working for the Department of Conservation and Recreation in the Division of Natural Heritage (DNH). He wrote up his findings while a doctoral student in the School of Forest Resources, College of Agricultural Sciences, Pennsylvania State University, and they appeared in the Virginia Society of Ornithology journal *The Raven* in 2001.

In 2004, I learned about David J. Horn, at Aurora and then Millikan universities in Illinois, who expressed interest in bird-window strikes. He was keen to address how to prevent them in his role as an ornithological scientist affiliated with the Wild Bird Feeding Industry (WBFI). David and his students continue to study this issue in Illinois.

Yigal Gelb and Nicole Delacretaz at New York City Audubon invited me to oversee a bird-window study in Manhattan from 2006 to 2007. Their work resulted in three important scientific publications that brought added attention to the issue in urban areas. Stephen B. Hager, an animal behavior and conservation biology professor at Augustana College in Rock Island, Illinois, became interested in the topic as early as 2004. His interest, in part, resulted from interacting with an enthusiastic undergraduate, Heidi Trudell, who collected window-kills and passionately tried to convince the leaders of her institution to prevent them at Principia College in Elsah, Illinois, just north of St. Louis. Steve and his colleagues and students have been publishing the results of several bird-window investigations since 2008.

In addition to publishing assessments of various bird-window collision prevention methods, I have written periodic review articles in 2006, 2009, 2010, 2014, and 2015 in peer-reviewed scientific journals and books, to update our collective knowledge and further stimulate interest in this conservation issue. Recent publications have increased credibility for the topic in the scientific community. These studies used sophisticated mathematical modeling tools to evaluate the most complete and objective data available on human-associated avian mortality figures. They appeared in print in 2013, authored by Craig S. Machtans and colleagues affiliated with Environment Canada, and in 2014 by Scott R. Loss and colleagues affiliated with the Migratory

Bird Center, Smithsonian Conservation Biology Institute, National Zoological Park and the U.S. Fish and Wildlife Service.

International scientific literature is also growing to alert and stimulate action globally. Notable principals and their beginning years of interest include Martin Rossler, from 2004 in Austria; Marco Dinetti and Giovanni Boano, from 2009 in Italy; Heiko Haupt, from 2011 in Germany; and from 2012 Erin Bayne in Canada, Adam Zbyryt in Poland, and Rose Marie Menacho Odio in Costa Rica. Others include, from 2015, Natalia Ocompo Penuela in Colombia and Ian MacGregor-Fors in Mexico; from 2017 Augusto Piratelli and Fabiano Montiani in Brazil; from 2018 Young Jun Kim in the Republic of Korea; from 2019 Natalia Rebolo-Ifran, Agustina di Virgilio, and Sergio A. Lambertucci in Argentina, and from 2020 Andreia Garces, Isabel Pires, Fernando Pacheco, Luis Sanches Fernandes, Vanessa Soeiro, Sara Loio, Justina Prada, Rui Cortes, and Felisbina Queiroga in Portugal.

Within the scientific community, spurred by what seemed to be a lack of meaningful interest and action, I also tried at annual professional meetings to interest and inform my colleagues about the hazards windows pose for birds. One special opportunity occurred at the 1998 North American Ornithological Conference held in St. Louis, Missouri. I titled one of two oral presentations at this meeting *Glass: a deadly conservation issue for birds*. My plan was to use this venue to ask my colleagues: "Given what has been published about the attrition of birds from striking windows, why, as professionals dedicated to birds, are we not individually and collectively alarmed, or minimally seriously concerned about windows as a conservation issue for birds and people?"

On purpose, I prepared no visual aids to distract the audience from the question. I did, however, expect to acquire some visual material at the meeting site by searching beneath the windows of the hotel or nearby buildings. This practice had always resulted in victims or victim remnants for use in my previous presentations on the subject. My search produced a strikingly attractive male Red-winged Blackbird in fresh plumage at the base of one of the hotel windows, facing the Mississippi River. I also discovered a Yellow-bellied Sapsucker beneath a large storefront window across the street from Bush Stadium, where the St. Louis Cardinals baseball team played. Prior to beginning my scheduled talk, I laid these specimens before the audience, and explained that, except for these window-killed specimens, I had no other visual aids. I emphasized that I was there simply to ask: "Given what we now collectively know about avian mortality from striking windows, why is there so little concern?" I followed with a brief review of available facts, and then asked if one or more of them knew something I did not. I repeated my plea for a reply, emphasizing that no one should be inhibited about speaking up. I vowed to welcome their views about why this issue seemed so underappreciated and how to fix that. I added that no one need worry about embarrassing me or have any other concerns about my personal feelings. Please, I repeated, offer me your knowledge, opinion, and advice, now or after you return home. The result: silence. The one exception was a former professor who afterwards commented that I should have used visual aids. Incredibly, I received no other response from the 70-plus attendees, at my presentation or thereafter.

Another memorable opportunity to stimulate interest and action from my colleagues occurred at the April 2005 joint annual meeting of the Association of Field Ornithologists and the Wilson Ornithological Society held at the Sheraton College Park Hotel in Beltsville, Maryland. This time I tried the tactic of using creative art. For some time I had been collecting newspaper and magazine cartoons where the subject involved birds and windows. Over the years, more than a few students have added to my collection by sending me comics that have come to their attention. Among notable cartoonists that have included bird-window collisions as a subject are Charles Schulz in his *Peanuts* series on at least two occasions, Bill Watterson in his *Calvin and Hobbes* series, Joe Martin in his *Cats with Hands* series, and Gary Larson in at least five of his *Far Side* cartoons. My meeting talk title was *A humorous look at a deadly conservation issue: birds and glass*. I used select cartoons to once again review what we knew about the issue and ask why they or others were not more concerned. Like the 1998 meeting, I was met with silence.

Although some colleagues and organizations, some quite prestigious and influential, were beginning to express concern, to this day the need remains to capture a critical mass of scientific colleagues, citizen scientists, other conservation-associated advocates, and most of all members of the public capable of voicing their concern to lawmakers and building industry professionals, asking to make the built environment safe for birds. The sentence with which I concluded my presentation is still relevant today: "... at least these cartoonists have taken this issue seriously." To be fair, my colleagues were interested enough to come to my talk, which was especially well attended, and my appeal was founded on the hope that they, more than I, had greater influence to advance this worthy conservation cause.

Bird-window collision cartoon from Peanuts, by Charles M. Schultz.

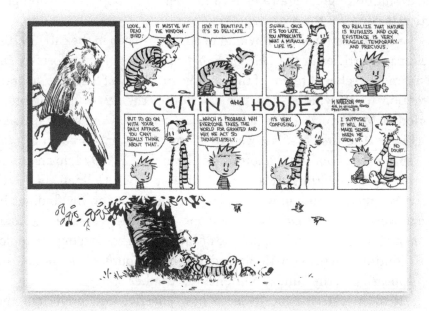

Calvin and Hobbes bird-window collision cartoon by Bill Watterson.

Chapter 8

Cats with Hands bird-window collision cartoon by Joe Martin.

Over time, scholarship did attract the attention of popular writers, and they in turn have educated a far broader audience than the scientific community. An account of some of these popular efforts will explain the evolution of greater awareness and offer some hope for meaningful action to come for this worthy cause.

Serious interest in bird-window collisions began in the early 1970s. Newspaper articles in the *Chicago Sun-Times* highlighted efforts by Dr. William J. Beecher, director of the Chicago Academy of Sciences, to draw attention to 30 window-kills per day during the 1974 fall migratory period at McCormick Place, a convention center on the shore of Lake Michigan. The first media coverage of my efforts appeared in a local newspaper, the *Southern Illinoisan*, on November 27, 1974, in an article written by bird enthusiast and outdoor writer Ben Gelman. In 1976, Chris Moenich, a student writer, provided an update on my findings in the university's *Daily Egyptian* newspaper. Over the years, Ben Gelman remained interested and supportive of my work in addressing the topic, and continued writing periodic updates in the *Southern Illinoisan*, the last appearing in the Sunday edition on December 6, 1992.

It took years for me to learn how effective and important the media are in educating. Experience has also taught me how challenging it would be for me to write about this topic in popular literature, in newspapers, magazines and, to a lesser extent, radio, television, and internet reporting.

My inexperience resulted in my missing a great opportunity to bring attention to this conservation cause. I had just accepted my first and only academic appointment at Muhlenberg College in Allentown, Pennsylvania, right out of my graduate school program. I have since long forgotten his name, but during my first semester teaching in the fall of 1979, a writer for *National Geographic* was on campus to give a series of lectures. During his stay, he heard about my bird-window research and came by my office, introduced himself, asked a few questions, and then shared that he thought the topic was especially interesting and he would like to write an article about it for *National Geographic*. I was not surprised by his interest, but I surely had an inflated view of my potential abilities to communicate about it. I vividly recall replying to him after a noticeable, bordering on uncomfortable, silence, after which I said, "Gee, I was hoping to do that." I do not recall his exact reply, but I am certain it was one of respectful acknowledgment, and we parted exchanging well wishes. I never heard from him again, and no others from *National Geographic* ever contacted me with an invitation to write an article for them on this or any other subject. Perhaps the future would have been kinder and more productive if I had accepted his offer to apply his talents and *National Geographic's* reputation on behalf of birds. The harsh lesson I learned was that the opportunity to write for a popular audience is a rare event for people like me without influential contacts. I vowed thereafter only to encourage any media source to apply their talents to teach more effectively than I could about this vital conservation cause.

In 1983 I presented a paper on bird-window collision prevention at the celebrated centennial meeting of the American Ornithologists' Union (AOU) at the American Museum of Natural History in New York. A few colleagues expressed serious interest and offered encouragement about the popular appeal of the issue. One was a member of the Natural History Unit of the British Broadcasting Corporation (BBC), Jeffery Boswall. He asked to interview me for a BBC radio segment as part of his coverage of the conference. Of course, I agreed, did the interview, and was informed that it aired during or shortly after the meeting. Thereafter, I enjoyed infrequent but always-friendly exchanges with Mr. Boswall, who showed continued interest in my studies. I arranged for

Chapter 8

him to visit our campus to describe his work producing wildlife film and video. I recall our students showing special interest when he described how some very renowned filmmakers artificially staged scenes to depict and imply natural behavior in natural settings.

Another early publishing opportunity that at first seemed promising also ended in disappointment. This exchange occurred with the National Audubon Society's magazine *Audubon*, and validated how important it is to be introduced personally to decision-makers rather than simply being told who to contact. In 1984, I was elected to the board of trustees of the prestigious local Hawk Mountain Sanctuary Association (HMS), which was of mutual benefit to my employer. At my first meeting, fellow board member Michael Harwood introduced himself by stating that he was very familiar with my research on birds and windows. He went on to describe how his editor at *Audubon* had asked him to buy my dissertation and prepare an article for the magazine. He explained that he did obtain it, but also how it remained on his shelf because he received more immediate assignments that kept him from preparing a bird-window piece. He was an enthusiastic advocate and recommended I write his contact at *Audubon*, executive editor Roxanna Sayre, introducing myself and telling her that he had endorsed the topic and suggested I write the article myself.

I recall being very excited thinking that I had a credible personal contact that would probably welcome a bird-window article, offering me an opportunity to write for a large, influential audience. I enclosed a copy of an article outline because I was so certain of her interest. Three months later, I received an unsigned printed postcard from "THE EDITORS" simply stating *Audubon* magazine "... cannot use an article like the one you suggest." This experience convinced me that lack of a direct contact, talent, or both, inhibited my opportunity to write about my cause in a popular medium with a large readership. The experience reinforced my vow to offer all the information and encouragement I could to media sources interested and willing to inform the public about birds and windows. And after that, I never failed to make a point of telling media contacts that they were the very best educators, and that I highly valued anything they produced to inform and help save more birds.

An invitation from a special-interest bird publication afforded me the chance to learn more about popular writing through trial and error. Rick Bonney, at the Cornell Lab of Ornithology, was one of those interested colleagues who introduced himself to me after my presentation at that 1983 AOU meeting in New York. He asked me to prepare a short article for the Cornell Lab's *Living Bird Quarterly* magazine. I eagerly agreed and promptly did so upon returning home. Within a few weeks, the manuscript I submitted with unlimited hope and pride was returned completely rewritten. I was at first stunned, then disappointed, and finally wondered whether I just did not possess the skill to write popular articles. But I refused to give up or have my passion for the cause diminished. After all, young as well as old dogs are known to learn new things. I carefully studied what was done to my original manuscript, and I rewrote what they rewrote, but in my own words. I then returned the manuscript to Mr. Bonney, asking him to consider my rewritten version. I insisted that if this offering was still unacceptable, they should use what was acceptable and replace my name with that of the actual author, because it was the message, not the author, that was of value. They accepted my rewrite, and the article appeared as I wrote it the second time.

Six years later, another Hawk Mountain Sanctuary Board member, William H. Thompson, the publisher of *Bird Watcher's Digest,* invited me to write an article for the cause. This time my original was accepted and appeared in 1992. Working with Mr. Thompson finally convinced me that, although we can all improve at whatever we do, at least in my case it was not the writing, but a belief in the issue and a sincere commitment on the part of decision makers that would result in opportunities to contribute to popular literature.

By 1994, my exchange with a freelance writer keenly interested in the bird-window issue further reinforced my conclusion that, like my scientific colleagues, editors of popular literature mostly were unaware and needed convincing that the topic was worthy for their publication. Joe Clark, a Canadian writer who felt the topic was so important it merited a large audience, asked to write a piece for *Popular Science*, a large-circulation magazine. I gave him my full cooperation, in keeping with my vow to encourage and help every media source. Unfortunately for us and the birds, the editor evaluating his proposal wasn't convinced the topic was "ripe" for publication, and among other reasons, said in her reply, "I am an avid birdwatcher, and I have never heard of this phenomenon in birding circles." Of course, her reply reinforced how some influential

Chapter 8

birders were either unaware or did not believe the evidence in my science journal articles merited their interest and action.

In 1997, I had a dramatic media opportunity with remarkable public exposure possibilities. Sharmila Choudhory called me from the natural history unit of the BBC, explaining that she was an assistant to Sir David Attenbourgh, who was engaged in producing a video series called *The Life of Birds*. She described how he planned to devote the final segment to bird conservation and was interested in including the bird-window collision issue. She was calling to find out where they could record birds crashing into windows. She explained they had the technology to simulate such an event but preferred to capture actual collisions. I recommended downtown Toronto, where volunteers from the Fatal Light Awareness Program (FLAP) collected large numbers of dead, dying, and injured window casualties during the fall and spring migratory periods. I knew they would be eager to help. When the final episode appeared, this fantastic opportunity had long passed. For reasons not shared with me, the series had no final conservation component.

Opportunities to advance this conservation cause have typically been unexpected, and such was the case with an uncanny coincidence that occurred in November 2001. Early that month, my wife, Renee, accompanied me to the Canadian side of Niagara Falls. I was there to represent my co-authors and present a paper on Armenian water birds at the annual meeting of the Waterbird Society. While we were enjoying the falls, looking to the U.S. side, we could see what looked like a very large covering draped over the cliff-face. We asked fellow tourists and the staff at the meeting hotel, "What are the Americans doing over there?" No one seemed to know.

About two weeks after returning home I received a call from Thomas Lyons, director of New York State Parks, Recreation, and Historic Preservation, who asked: Ever been to Niagara Falls? My reply: Funny you should ask, just recently. He then explained how the observation tower at the New York State Park at Niagara Falls had deteriorated. When a plan to remove it entirely was announced, a public outcry ensued, and the state government eventually agreed to replace it. The original design for the replacement called for covering the entire tower with mirrored glass, with the intent that it would blend into the surrounding environment and in effect be invisible. He went on to share that almost immediately after the design plans were revealed, another cry went out that the entire area of which the park is a part is an officially designated Important Bird Area (IBA), and reflective glass is a documented bird killer. The critics argued that, rather than building an aesthetically pleasing substitute, the park system was planning to create a horrific bird killer at a very important site where birds congregate in large numbers and people come to see them. He went on to describe how, after learning about my studies, he had been looking for me for about a year and had only recently obtained my contact information. He invited me to review and comment on the environmental impact of the new tower construction. He also asked me for recommendations on preventing bird kills at what glass was proposed for the design. Of course, I eagerly agreed and expressed my gratitude for answering our question about what the Americans were doing covering part of their side of Niagara Falls.

I later learned that they used my suggestions in designing the new tower, and the outcome, based on monitoring, was a success. I did not actually see the new tower until ten years later when, in 2010, I was invited to a meeting about birds and windows in Toronto. My route there permitted a visit to the observation tower. I found its glass walls had prominent white stripes, separated by the spaces I had recommended. I was pleased and surprised to find a prominently placed sign at the entrance explaining the collision issue and how the tower's design was protecting birds. Less attractive to me, the text of the sign prominently acknowledged several personalities from the Cornell Lab of Ornithology for photographs and bird information used in the display, among them the renowned ornithologist Olin Sewall Pettingill, Jr. Other images from independent photographers and the U.S. Fish and Wildlife Service were depicted, but no mention of the source of the science that determined the design. My personal disappointment likely had a tinge of vanity, but this surely qualified as more evidence supporting the phrase one newspaper journalist had used about me, calling me the Rodney Dangerfield of ornithology, harking back to the comedian who drew laughs by complaining that "I don't get no respect." More seriously, I knew my employer would have appreciated being acknowledged as the source of the research used to create the bird-protecting design.

Surprises. It has been my experience that most surprises are not good. The hot water that does not show up for a morning shower, the vehicle that does not start, and the expected easy household need that balloons into another mortgage. No matter how infrequent or rare, good surprises do occur, even outstanding ones, such as news of an awaited and hoped-for pregnancy. One of my many good surprises came when the State Park at Niagara Falls, our nation's oldest state park, agreed to include a few lines acknowledging my employer and my contribution to the design of the observation tower, no matter that it was 16 years after they replaced it in 2001. The surprise was that they agreed to do it at all.

What changed their mind was the extraordinary effort to increase publicity for my employer, Muhlenberg College. A recently appointed new president, John I. Williams, Jr., was seeking ways to inform the world about the value of liberal education, and understandably and specifically, the special value of doing so at Muhlenberg. He saw my design for the observation tower as a practical contribution, of the kind regularly made by liberal arts institutions but too frequently overlooked. He identified this particular contribution to a practical problem by those engaged in liberal education to be especially appealing and noteworthy. Most informed people associate liberal arts teaching about worldly subjects and emphasizing the use of critical reasoning. An unkind and inaccurate description is that such education introduces students to everything, but nothing specifically. The use of my research at Niagara Falls was, to him, an example of how thinkers from liberal arts institutions can solve practical problems.

Observation tower at New York State Park at Niagara Falls showing striped pattern to prevent bird-window collisions.

The Kiki Network, owned and operated by Kiki Keating and staffed with her talented assistant Erica Finkelstein, was hired by President Williams to promote and increase the national and international visibility of Muhlenberg College. She began by conducting interviews and immediately found publicity value in my connection to Niagara Falls. She was so taken by it that she made it her personal goal to have Muhlenberg College added to the sign explaining the birds and windows issue at the base of the observation tower. In its original form, the sign explained that the stripes on the tower were there to prevent bird-window collisions. The new sign does the same, but adds a few more lines crediting the design to research conducted at our liberal arts college. Thank you, Kiki and Erica, for a good surprise, and thank you, Niagara Falls State Park, for agreeing to change the sign and for hosting the December 12, 2016, ceremony acknowledging Muhlenberg College among its many efforts to educate the 8 million to 9 million annual visitors to this iconic geological wonder.

By 2003, enough public exposure had occurred to attract the attention of Maryalice Yakutchik, who teaches journalism in the Department of Communications at Loyola College in Baltimore, Maryland. She discovered the bird-window collision topic and judged that the researcher studying it was overlooked, which added a human-interest component to a potential story. She introduced the subject and my role to her class, and asked them to assess whether there was a story worth writing about. They gave her the go-ahead, and, having written about a previous professor for the *Philadelphia Inquirer* Sunday magazine, she approached the paper with the prospect. They agreed. Her article was titled *Fatal reflections*, and it appeared in the *Inquirer Magazine* on May 11, 2003. It summarized the lethal threat of windows for birds and flattered me by describing my work as especially important and respectable. It also highlighted Swarthmore College, which had committed to putting uniquely designed bird-friendly sheet glass in its proposed new science building.

In February 2004, likely prompted by the *Inquirer Magazine* article, Joann Liviglio, a Philadelphia-based reporter for The Associated Press, wrote an article about birds and windows that was distributed over the wire service. Some 400 newspapers and a few television stations carried the story. It also appeared on the CNN website, which prompted a telephone call from an individual who introduced himself as a neighbor. He told me he'd seen the bird-window piece on his computer at work and explained that he worked in an all-glass building dedicated to communication technology, along a major highway between Allentown and Bethlehem, Pennsylvania. I was very familiar with the building. I frequently passed it on travels to and from my night and summer jobs teaching about birds at Moravian College in Bethlehem. I explained how watching the construction and completion of his workplace had me predicting that this building, like countless other glass-covered structures I see in my travels, would be another bird-killer, and by the looks of it, an extravagantly effective one. The computer work they did in this building and the security it demanded kept me from searching the area beneath its glass walls in person. My caller, however, gave me the opportunity to ask if he ever witnessed bird strikes. Without hesitation, he replied: "Oh my, yes, frequently, from birds the size of a Canada Goose to sparrows."

I received a mountain of email contacts from citizens who read Joann's AP article on the topic that appeared nationwide. They offered volumes of eyewitness accounts of birds hitting windows and recommending how to stop it. It was a physical challenge, but I politely replied to every one of the hundreds of emails, except two who were vulgar and insulting critics with human-centered interests. I did not learn anything new from the legions trying to help both the birds and me. For their well-meaning efforts, I thanked and expressed my gratitude to each of them for their concern, encouragement, and especially their opinion that the matter needed meaningful attention, which I invited each to consider informing others about and taking other meaningful action.

By far the most extensive major coverage up to that time occurred with the publication of a feature article on birds and windows titled *Clear & Present Danger* in the March 2004 issue of *Audubon* magazine, by David Malakoff. The article resulted from my attending the National Audubon's annual meeting in Los Angles the previous fall. To assure immediate publication, the editors at *Audubon* requested a major gift, specifically suggesting that I ask my long-time research benefactor, Sarkis Acopian. Considering the importance and potential impact of this educational opportunity, I ignored my revulsion to asking others for money no matter how worthy the cause. I attribute this inability to ask others for financial gifts to worthy causes to having been raised by caring people who struggled to pay for necessities and did not comprehend philanthropy, a luxury they could not practice. But in this case, I asked. Mr. Acopian kindly gave the $25,000 requested, and the article appeared in the next issue.

In November 2005, John Nielson, a science reporter for National Public Radio (NPR), contacted me, asking if I could contribute to a story he was planning about birds and glass. He was attracted to the topic by the commitment of Swarthmore College to use bird-friendly glass in their new $71-million science building. There was a documented history of bird kills from striking windows on their campus. The records were collected by faculty member and ornithologist Timothy C. Williams, who described to me the extirpation of nesting Ruby-throated Hummingbirds from college grounds, attrition he attributed at least in part to window strikes. After learning about these deaths, the college's green committee made bird-friendly glass a priority for the new building. The NPR segment appeared that December on the program *Morning Edition*.

Remarkably, Anthony (Tony) Brian Port, a birder and lead scientist for the external film manufacturer CPFilms, Inc., heard the *Morning Edition* segment. He already was aware of the issue. Some of the films his company put on glass transformed them into killers by offering a mirror image of facing habitat and sky. Birds hit these reflective illusions with fatal consequences. He called me to discuss ways his company could be party to resolving the threat. His affable manner assured we would quickly become collaborators and friends. He invited me to his company's headquarters in Martinsville, Virginia, to consider a formal collaboration. With other CPFilms principals, we discussed options, especially those using ultraviolet (UV) signals. We continue an ongoing exchange, with the goal of bringing what I believe is the most appealing and elegant solution to preventing bird-window collisions, using protective UV signals that birds see and we humans do not.

I received my first contact about bird-window collisions in the Chicago area in the fall of 2002. Libby Hill, of the Evanston North Shore Bird Club, just north of Chicago, wrote to me about bird kills on the campus of Northwestern University. After an engaging exchange of information, Libby invited me to visit in the fall of 2004, to meet with Northwestern staff to address the issue. She also scheduled me to make a presentation to her club. Several attendees expressed special interest, including Donnie Dann, a legend in environmental awareness. He offers sage advice on a wide-range of environmental issues and best practices in his periodic newsletters. I also met Randi Doeker, president of the Chicago Ornithological Society. Collaborating with the Chicago

Chapter 8

Department of the Environment and other Chicago government principals, Randi became the principal organizer of a conference in the spring of 2005 titled *Birds and Buildings: Creating a Safer Environment*. The conference brought together architects and conservation leaders from around the U.S. to address the bird-window issue. We met on the campus of the Illinois Institute of Technology (IIT), where sheet glass figures prominently in architectural education and design. Immediately following this event, Randi created a website titled *Birds and Buildings Forum* to share information among all interested constituencies and to reach the widest possible audience.

While in the Windy City, I met Robbie Lynn Hunsinger, a classical musician who had developed a passion for saving birds from windows. She was the founder of the Chicago Bird Collision Monitors, with whom I had been corresponding since 2003. Robbie and her successor, Annette Prince, established routes in downtown Chicago to collect the dead, dying, and injured casualties. Their monitoring effort was modeled after what Michael Mesure and his co-founder Carolyn Parker, founders of FLAP, began in Toronto in 1992, when they attracted a dedicated volunteer group to patrol prominent collision sites in the Greater Toronto Area. FLAP was based on the initial interpretation that most collisions occurred at night, when spring and fall migrating birds were attracted by city lights burning in high-rise buildings. It is fair to claim that I played a key role in convincing FLAP that few birds actually die from striking buildings at night. More detailed study revealed that most kills occur in the early morning hours, when migrants are in the urban canyons of concrete and glass. Victims are deceived by clear and reflective windows while seeking food and shelter. At the end of their nighttime journey, migrating birds return to the ground to rest and feed. If the weather is clear, migrants typically fly high enough away from the influence of city lights. During inclement weather, when cloud cover is low, birds on migratory passage fly low enough to be attracted by urban lighting, especially from skyscrapers. The attracted migrants become confused, flying in and out of the lights, until they are exhausted. They then flutter into the threatening city landscape below.

Following FLAP's lead and copying their practices, the Chicago Bird Collision Monitors continue to document window casualties, as do, among others, committed programs organized by Rebekah Creshkoff in New York City, Wendy Olson in Baltimore, and Anne Lewis in Washington, D.C.

During my first trip to Evanston that fall of 2004, Randi took me on a few routes Robbie and Annette monitored. I was amazed at the number of Brown Creepers and Lincoln Sparrows we found as window casualties. She showed me the famous McCormick Place, a convention center with walls of glass facing Lake Michigan, where 200-plus victims were killed daily during migration periods. Adjacent to the convention center was the infamous Soldier Field, home of the Chicago Bears. The NFL stadium was of special interest to me because a few years before, I had been contacted by an architect engaged in renovating the site. He asked for my ideas about preventing bird strikes at stadium windows and shared that it was too bad I had not been in Chicago earlier that year for the annual meeting of the American Institute of Architects (AIA). He explained that the theme of this year's AIA meeting was the environment, and my studies seemed especially relevant. I shared in turn that I tried to attend. I explained that the National Audubon Society (NAS) had asked me to submit a request to speak about the window issue for birds, but I had been turned down. Encouraged to apply again by the NAS, I did so the following year, but my application was again unacceptable. Although no explanations were offered, I suspected that, after feedback from a member of the program selection committee, the AIA annual meeting organizers viewed the topic as uncomplementary to their professional interests. I lamented to my advice-seeking architect caller and also later to an AIA program selection committee member that if given the chance to speak to their colleagues, I would have enthusiastically sought their collaboration, emphasizing how the birds could use and needed their creative talents to address these unintended and unwanted tragedies.

Engaging these talented building professionals with this message is a vital goal. Walker Glass of Montreal, Quebec, Canada, invited me to speak at an information meet-and-greet just prior to the 2019 AIA annual meeting in Las Vegas, Nevada. An attendee introduced himself to me after my presentation and told me he was on the evaluation and selection committee for the AIA program. He said he believed the topic was essential, or should be, to AIA members, and strongly encouraged me to apply to speak at the 2020 meeting in Los Angeles. He followed up later with an email further encouraging me to apply. I promptly did, and about a month later received a notice that my request was duly considered but not accepted. Even with the critical need to continue to educate this extremely important constituency, my three strikes with the AIA suggest a new approach and more persistence is required.

Also adjacent to McCormick Place is the Chicago Field Museum, whose collection manager, David Willard, and conservation ecologist Douglas F. Stotz had been systematically collecting window-killed birds at the convention center. Their recommendation to manage lighting at the center, turning off the lights at key times, resulted in an 80% decrease in strike fatalities. A city-wide lights-out program for Chicago was initiated by Judy Pollock of the Illinois Chapter, National Audubon Society. I learned from her that the effort to turn off lights was having success among building managers. We humans on behalf of birds can thank lights-out programs in those cities where they occur across North America for saving countless bird lives during fall and spring migratory periods.

During the 2005 Chicago conference, I met a number of interested and committed architects, all AIA members, who sought to design bird-friendly buildings. Among them were the faculty and students at the Illinois Institute of Technology (IIT), on whose campus birds were killed striking the sheet glass in several buildings and at clear noise barriers. I also met Ellen Dineen Grimes, a professor in the School of Architecture, University of Illinois at Chicago, who expressed special interest in the topic. I met Jeanne Gang and Mark Schendel, the principals of Studio Gang Architects, located in downtown Chicago. They and their staff at Studio Gang would leave a lasting impression on me. Soon after the conference, they invited me to comment on their bid to design the proposed Ford Calumet Environmental Center, a project initiated by the City of Chicago Department of Environment and the State of Illinois to redevelop an abandoned, highly polluted site in Calumet, just south of Chicago. Claiming I contributed meaningfully to their submission, Studio Gang was selected as the winning entry in the international design competition. Their plans included using discarded materials and bird-friendly construction. Two years later, in 2007, I accepted an invitation to visit their studio and further collaborate on the Calumet Center and other projects where they planned to incorporate bird-friendly designs.

The American Bird Conservancy (ABC) is a committed and effective avian conservation organization. I first learned of their interest in bird-window collisions when they invited Steven Price of the World Wildlife Fund Canada (WWF) to speak about the topic at their policy council meeting in March 1997. Steven Price and Michael Mesure of FLAP had partnered and recently published *Collision Course: The Hazards of Lighted Structures and Windows to Migrating Birds*. Other than this initial concern about birds and windows, ABC was mostly silent about the issue while principally dedicated to alerting the world about the dangers of communication towers. In November 2003, ABC returned to the topic of birds and windows and invited me to speak at their policy council meeting in Washington, D.C. I am certain that my invite was due to Alicia Francis Craig, now Alicia Francis King, whom I first met at an ornithological meeting in the late 1990s, when she was working for Wild Birds Unlimited. In 2003, she was serving as the director of the Bird Conservation Alliance at ABC and planned to highlight the window issue on their website, similar to what she had done at Wild Birds Unlimited. In October 2007, ABC took it a step further, committing to a specific staff position called the Bird Collisions Campaign Manager. Karen Imparato Cotton was appointed in early 2008. I knew Karen from her days working at New York City Audubon, where she sought and enlisted my help trying to obtain grants to further study birds and windows and how to prevent collisions. I enjoyed a warm, especially friendly, and respectful relationship with Karen, from our first meeting and after she began her work with ABC. We talked regularly, at least every week. Most of her initial efforts involved trying to incorporate bird-safe glass as an evaluation subject in the United States Green Building Council's (USGBC) Leadership in Energy and Environmental Design (LEED) program. LEED assesses how environmentally-friendly buildings are by grading various structural features. It is an incentive for architects to encourage responsible environmental designs. Almost too sad to share, her work on behalf of birds came to an end along with everything else when a brain tumor claimed her life in July 2009. Christine Sheppard, an ornithologist and former curator of birds at the Bronx Zoo, succeeded her.

An extremely important contribution by ABC occurred at a 2009 conference about birds and windows organized by FLAP in Toronto. At that meeting, Chris volunteered ABC to establish a Collision Listserve on the World Wide Web. This initiative was and continues to be essential. The Collision Listserve informs all interested parties about the latest developments and provides a means for those who have questions to obtain the best answers from the collective wisdom of all members.

Chapter 8

In December after the 2005 Chicago conference, executive director E. J. McAdams and colleagues at New York City Audubon convened a Bird Safe Glass Working Group. The purpose of the Manhattan meeting, which included Karen Cotton, was to bring selected people together to consider what to do next to protect birds from windows. I was invited to make a presentation on the results of my latest investigations, but more important for me was another presenter whose reputation was well known to anyone interested in government's role in dealing with human-associated bird kills: Albert M. Manville, II. His formal work title was Senior Wildlife Biologist and Avian-Structural Lead, Division of Migratory Bird Management, United States Fish and Wildlife Service.

My riveted interest in him was his past description of the USFWS's responsibility of bird protection, which he emphatically and dramatically repeated during his presentation this day. He stated that by authority of international treaties with Mexico, Japan, Canada, and Russia, known in the U.S. as the Migratory Bird Treaty Act (MBTA) of 1918, as amended, the USFWS is obligated to prosecute the killing of a single individual bird protected under this legislation. Except for a very few introduced species, almost all of the 700-plus North American species of birds are protected under the MBTA. Al further emphasized that the killing did not have to be intentional. Violators may be intentionally or unintentionally involved in the killing of the protected species. This includes loss of life from diverse human activities such as pesticide applications, collisions with power lines, communication towers, wind turbines, and sheet glass windows. If convicted, violators could be fined and even imprisoned. I was aware of efforts by Al and others to prevent the unintentional killing of protected birds from power lines. This was a success story resulting from a collaborative working group that included experts and authorities among academics, government scientists and administrators, and principals of the electrical power and utility industry. But until this meeting, Al and his colleagues at USFWS were silent about the unintentional bird kills occurring at residential and commercial building windows. For well over two decades, my studies had documented the astronomical threat windows pose to birds compared to the far more meager threat of other human-associated mortality, including power lines, that appeared in several peer-reviewed scientific journals. As Al spoke about the glass threat, my frustration mounted when he repeated, on more than one occasion, how important it was to have peer-reviewed scientific papers documenting the glass threat to birds. Only with such documentation could the USFWS justify action toward this type of unintended killing. What Al described as needed was already present in peer-reviewed publications. So, did he and others at USFWS not believe or respect my peer-reviewed published results?

At a break between speakers, I had my chance to ask Al this question and more. Why has the USFWS taken no action about window-kills under the MBTA, given the criteria he just described to justify it? Were there flaws in my results, analyses, or interpretations? I asked for his patience in replying, and added that I recognized the challenge USFWS had in enforcing this issue. Given the nature of the hazard, every homeowner, commercial property manager, government agency and school housed in a building with windows is liable and likely a violator of bird protection laws. I also acknowledged that as a homeowner aware of lethal bird strikes at my own windows, I and likely a majority of U.S. citizens are violators. Furthermore, any reasonable person familiar with this type of mortality would think it ludicrous that the USFWS would be expected to arrest every culpable citizen for birds unintentionally killed at their property. Even so, I believed it reasonable for USFWS to enforce bird protection laws at sites like McCormick Place in Chicago, where 200-plus birds die flying into windows in a single day. Al politely heard me out and surely sensed my frustration not to receive USFWS support for this cause. He said he had read and found my research compelling. He acknowledged that the work was published in credible scientific sources. He then attributed no action to date by the USFWS due to limitations, most importantly those associated with enforcing the MBTA, having complex legal implications that render it ineffective for this and other wildlife protection measures. The USFWS use of MBTA in what successes they obtained was to encourage voluntary cooperation from prospective violators, given the uncertainty of court outcomes if direct enforcement was used. Another huge limitation Al described was a lack of time and personnel to dedicate to this specific issue. Another was the overwhelming nature of the problem. The USFWS was stymied about how to address unintended window-kills, given that they literally happen everywhere. Between this conversation and his official retirement from government service in 2014, Al was a leading advocate within the

USFWS to address avian mortality resulting from window strikes. His culminating effort as a retiring government employee was as a principal contributor to the USFWS's January 2016 publication *Reducing Bird Collisions with Buildings and Building Glass Best Practices*.

A literal striking outcome I attribute to the media coverage of the 2005 Bird-Safe Glass Working Group and New York City Audubon efforts to protect birds from windows was a noteworthy piece of art, a sculpture. Cai Guo-Qiang, an internationally renowned Chinese-born artist, created an exhibition in 2006 for the Iris and B. Gerald Cantor Roof Garden at the Metropolitan Museum of Art. Four works made up the "Cai Guo-Qiang on the Roof: Transparent Monument" exhibition, described as inspired by the open-air space atop the Lila Acheson Wallace Wing, which offers stunning views of Central Park and the Manhattan skyline. Among the four works was a 15-foot-tall clear glass pane called "Transparent Monument" that included replicas of dead birds at the base of the glass wall. In contrast to the well-intentioned motive to exhibit and teach about the lethal hazards that sheet glass poses for birds at this iconic site, was the obvious lack of awareness by the artist and the museum about the need to protect local free-flying birds from the sculpture. A wide enough netting enclosing the glass would have allowed for viewing and admiring the art without causing the collision deaths of real birds, actual victims that hit the "Transparent Monument" while it was on display between April and October.

I trace serious attention given to window-kills and their effects on bird populations, from scholarly to popular literature and other media coverage, to my relationship with the renowned Canadian ornithologist David M. Bird. My contact with David followed an announcement in 2000 about a bird–human interaction symposium at the joint meeting of the American Ornithologists' Union (AOU), British Ornithologists' Union (BOU), and the Canadian Society of Ornithology (CSO), to be held at Memorial University, St. John's, Newfoundland. I was certain the bird-window issue was more than appropriate and should be part of this symposium. David, a well-known ornithologist from McGill University in Montreal, was a principal organizer, and I pleaded my case with him. He agreed, and I received a coveted invitation that included the unforgettable benefit of seeing the diverse splendor of Newfoundland. After this meeting, David became a regular supporter and promoter, helping to educate about birds and windows. He highlighted the topic whenever new findings warranted coverage in his regular column about birds and natural history in the *Montreal Gazette* newspaper and *Bird Watcher's Digest* magazine.

One extremely important outcome of this growing body of technical and popular literature and other media coverage, to be described more fully later, is the interest expressed by glass and external film manufacturers in developing bird-safe products. These are products that need to be installed the world over to protect the world's birds. Glass manufacturers showing serious interest include Arnold Glas in Germany, AGC of North America, Goldray and Walker Glass in Canada, Pilkington of Nippon Sheet Glass (NSG) in the U.S. and U.K., and Vitro from Mexico and formerly Pittsburgh Plate Glass (PPG) in the U.S. External film manufactures include the Convenience Group in Canada, 3M, CollidEscape, Eastman Chemical, Erickson International, and Surfacecareusa in the U.S.

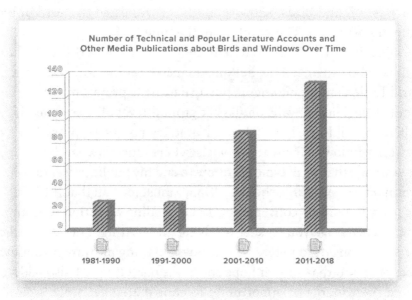

Despite all the validated scientific evidence and these apparent successes in getting the word out, this issue is still largely unknown or underappreciated as a serious environmental problem among the general public. Although a growing number of studies have revealed new details, there have been no revelations of any errors or discrepancies in the fundamental results of the earliest comprehensive investigation in the 1970s. Studies have consistently documented and confirmed the lethal threat sheet glass and plastic pose to birds in general, worldwide.

I offer two final stories to stimulate learning and action to protect birds from windows.

Disbelief hampers recognition of this threat, often by people who should know better, people who consider themselves environmentalists and identify with the animal welfare, conservation, legal, and building industry communities. At a 2005 meeting in Toronto, Joe Bartell, a tireless advocate for saving birds from windows in the Detroit area, described how, after moving to Denver, he asked the local Audubon Society what they were doing about birds killed flying into windows. Unanimously, they assured him that no birds die in this manner in Denver. They suggested that perhaps weather patterns or migratory behavior explained their fortunate immunity. Aware of my claim that we should expect bird kills whenever and wherever birds and glass coexist, he called a number of commercial window-washers serving the greater Denver area, asking if they ever found birds below windows, or witnessed or heard of birds flying into windows. Their reply? "All the time."

Rebekah Creshkoff is a citizen-scientist and New York City Audubon member who convinced the building managers at the now fallen original World Trade Center buildings to put up netting to protect migrating birds from the unyielding walls of glass at ground level of these magnificent multistory buildings. Rebekah is still at it, trying to save all the birds she can from windows in New York and elsewhere. But it is what she told me a few years ago that I still typically quote to end my public presentations. In a warm but serious tone, she said: "Dan, I must tell you that I still believe this is an important conservation issue needing attention. Certainly, I want to encourage you to keep studying and writing, but I also believe you will be long dead before anyone does anything meaningful about it." I always add to that: "I still hope she is wrong." Of late, I also add: "but for me the window of opportunity to contribute effectively is closing. For those of you reading this with the time, energy, and willingness to be party to saving more birds from windows, please accept my plea to learn, inform, and take every action appropriate for you to encourage bird-safe building practices whenever and wherever opportunity allows. Please join me and the growing number of others committed to this cause, if need be, ideally carrying on long after I am not able."

Chapter 9

Solutions – Solving the Problem

There are effective ways to protect birds from clear and reflective windows, and the means to do so are growing. Not so when I began my studies in the early 1970s. At that time, I vividly recall being advised by several prominent avian conservationists that if whatever I recommended to protect birds from glass interfered with the way they and others see out their windows, I would be ignored; or as they put it, I would lose!

To be effective, the solution must physically keep birds from glass or visually transform the space windows occupy into a barrier that birds will see and avoid. An introduction to bird study typically describes sight and sound as the dominant avian senses. Because humans also share these dominant senses, our mutual perception explains why birds are especially attractive to us. Our common perception offers clues about how to make windows safe. In an August 2016 Master of Science (M.S.) thesis, Nicole Marie Ingrassia at the College of William and Mary found evidence that sound can influence avian flight in ways that deter them from striking solid objects. As promising as sound may be, in my view, visual cues are likely to be universally more practical and effective than sound to alert birds to windows.

As previously shown, lighting, vegetation, water, fruiting shrubs and trees, bird feeders and other foods within the vicinity of windows can attract more birds into the lethal danger zone. But there are ways to mitigate this danger. Covering or orienting lights downward will limit their ability to attract birds. Placing indoor vegetation so that it is not visible to birds looking through a clear pane from outside will prevent an attempt to reach it. Locating attractants within one meter (three feet or less) of the outside window surface will prevent fatal strikes. My experiments with my students have documented that the closer an attractant is to the glass surface, the less risk to bird life. Although this seems counterintuitive, the protection results from birds coming and going to and from the attractant. As they leave a nearby attractant such as a feeder, even if they brush up against or hit the glass, the contact does not have enough momentum to result in injury or fatality.

My studies have repeatedly revealed that beyond one meter (3.3 ft), there is really no safe distance to place an attractant away from the window surface. Many writers addressing the issue of where to place feeding stations near windows have wrongly interpreted the results of our feeder placement experiment to mean it is safe to place attractants greater than 10 m (33 ft) from a window surface. In designing the experiment, I placed a 10 m limit on feeder placement, thinking that this distance would be the farthest acceptable to most homeowners or wildlife park visitor center managers, because anything greater would not bring the visiting birds close enough to be seen. What our experiments and countless other observations reveal is that, except for within one meter (3.3 ft) of the glass surface, placing attractants at distances greater than 10 m (33 ft) is no safer than closer locations, as long as birds are attracted to where windows occur. Birds are at risk of a lethal strike if they can enter the danger zone from any distance from the window surface.

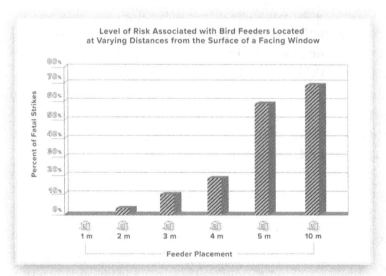

Birds are protected from windows covered by common mosquito netting (see www.birdscreen.com), or other fine mesh netting such as the kind used to cover vineyard grapes and other garden fruits and crops. Birds that are deceived and strike these barriers are protected by bouncing off and not hitting with enough force to cause them any harm. A successful architectural application preventing bird strikes is to cover the exterior of windows with sun-shades, such as louvers or other coverings offering the appearance of an opaque barrier or a perch just in front of the glass surface. Window coverings of this type have proven to be effective on the New York Times building in Manhattan.

Bird feeder placed near the glass surface of house window and university building.

Angled windows in the Visitor Center of the Eastern Shore of Virginia National Wildlife Refuge, Cape Charles, Virginia. Photograph by Robert Leffel, Deputy Refuge Manager.

Installing windows at a 20° to 40° angle, inward at the base from vertical, is another architectural design that was thought to protect birds from building up enough momentum to result in a fatal strike. My original expectation, and source of experimental hypothesis, was that window angling would protect potential victims because birds were less likely to fly into a reflected image of the ground than one of the facing habitat and sky. In 2004, my students and I found and reported in the journal *Wilson Bulletin* no support for this hypothesis. Birds were just as likely to fly into an image reflecting the ground as one reflecting what was above ground. We explained our findings by judging that the deflection of the strike off the angled pane lowered the force with which birds hit. The reduced force saved their life, and not an aversion to flying into the ground, which most, if not all, birds are accustomed to landing on. Angling windows will save the lives of birds that hit them, but it will not prevent or eliminate strikes.

As described earlier, I used both tunnel testing and field experiments to evaluate various means of preventing collisions between birds and windows. Tunnel testing was used to evaluate many preventive methods in a timely manner. Field experiments were used to confirm the effectiveness of preventive methods because this design accurately simulated windows installed in actual buildings. Accounting for controls, our current field design can test as many as seven methods over 30–90 days. It is unquestionable in my mind that a field experiment is required to obtain a confident evaluation of any preventive method. Almost all tunnel and field tests provide similar results, but when differences occur, they result from the many variables that limit tunnels from being able to simulate windows installed in actual human structures.

It is disturbing to some, myself included, that birds sometimes die in my field experiments. For me, these unfortunate and tragic losses are justified by the accurate results that no other experimental protocol can provide and the billions of bird lives that will be saved from these sacrifices. The results of tunnel testing do not simulate bird-window interactions as they occur in real human structures. By comparison, field experiments provide the most accurate and objective information about bird-window interactions because they create the exact conditions that occur in real human dwellings.

Writing in 2019, in the journal *Global Ecology and Conservation*, Christine Sheppard summarized the tunnel test results she and her colleagues carried out over several years, and unfortunately misrepresented my field experiments and their impact on birds associated with those experiments. She states "... Klem's subjects are often injured or killed ..." and "This may produce a negative impact on the local bird population." These assertions are without foundation and minimally misrepresent the essential value of conducting field experiments to confirm the true effectiveness of bird-window collision deterrent methods. In fact, the number of birds killed in my field experiments ranges from 2 to 22 (3% to 22%) of the recorded strikes. All lethal collisions are documented in all of my scientific papers. There is no evidence whatsoever that local bird populations are negatively affected at or around our field site. All field experiments are conducted during the non-breeding season to prevent the possibility of local bird casualties. Additionally, we have records from homes literally within a stone's-throw of our experimental windows that kill exponentially more birds than the relatively few fatalities that occur during our experimenting. The level of such residential casualties are common and documented by many peer-reviewed scientific journal articles, mine among them. We need field experimental testing as the most accurate means to assess bird-window deterrence methods, and as an unequaled, reliable tool to confidently validate those methods.

Recall that the surfaces of multipaned windows conventionally are numbered from the outside inward. Surface #1 is exposed to the outside environment and Surface #2 is the inside surface of this first pane. Surface #3 is on the next adjacent pane and Surface #4 is the opposite inside surface. The surface numbers increase in the same manner with each additional pane. A window with three separate sheet glass panels will have six numbered surfaces, and typically air or gas is what separates one panel from the other in a window composite. In all but those locations where habitat is visible behind clear panes, applying visible patterning to Surface #1 is the most effective means to alert birds to the invisible danger of windows. See-through effects are exceptions because the patterning can be applied on any surface of a multi-pane window and still be seen and alert a bird to avoid that space.

Alternatively, almost all windows, even perfectly clear panes, will act like a mirror, reflecting the facing habitat and sky when outside lighting is more intense than the darker indoor space the window covers. With reflective windows, patterns placed on any window surface other than Surface #1 are at risk of being hidden from birds looking at an illusion of the facing environment. In 2010, my long-time research assistant and colleague, Peter G. Saenger, noticed that the results of a series of field experiments revealed that even a clear film applied to Surface #1 created enough of a distortion to reduce window strikes by 50% compared to an unaltered control. Our field testing repeatedly validated how essential it is to apply preventive measures to Surface #1 to make windows safe for birds.

It is helpful to categorize preventive measures as short- or long-term solutions. Short-term solutions involve retrofitting existing problem windows. This is no small task, and manufacturers of retrofit methods should be encouraged, because offending windows exist in just about every human structure the world over. Long-term solutions refer to creative new sheet glass that is bird-safe for installation in remodeled and new construction, ensuring no bird fatalities will occur during the lifetime of the building.

See-through effect at a corridor between two buildings where habitat is visible behind one or more clear panes.

Reflective effect where facing habitat and sky is reflected off a clear or tinted panes.

My very first experiments addressing bird-window strike prevention revealed that the elements that make up a pattern applied to a window can be any shape or size. What is crucial is that the pattern must uniformly cover the entire window. The pattern elements must not be separated by more than 5 cm (2 inches) if oriented in horizontal rows, or by 10 cm (4 inches) if oriented in vertical columns. The spacing between pattern elements has been called the "2 x 4 Rule" or "Handprint Rule," the latter created by Randi Doeker. These patterns eliminate nearly all fatal bird-window collisions. Greater spacing between pattern elements increases the risk of a strike and casualties. Writing in the journal *Biologia* in 2015, Martin Rossler and his colleagues from Austria remind us that the use of "visual noise" on windows will be more acceptable to the public, private and commercial users, when pattern elements have maximum deterrence with minimum coverage. Using tunnel testing, they conducted 1,428 experiments evaluating various patterns and found the orientation, spacing, dimensions, and contrast of pattern elements determined their effectiveness in preventing strikes. Vertical stripes that are only 2 mm (0.08 in) wide separated by 10 cm (4 in) effectively prevent bird-window collisions. This vertical striped pattern covers only 7% of the window surface, providing a minimal amount of obstruction for human viewers. They reported that these fine lines were as effective as 13 mm (0.51 in) wide lines covering 50% of the glass surface. The study also documents that more contrast between window markings and the glass surface increase visibility and protection for birds.

Clear window with mirror reflection.

My field experiments support the spacing of pattern elements that Rossler and his co-authors and Sheppard reported. Caution, however, is still warranted about other features attributable to prevention until a field test can confirm the claims. As efficient and accommodating as testing is in tunnels, there is always the possibility of error caused by flaws in the test design. Some limitations that might compromise tunnel-testing results include the use of netting in front of the windows being tested that subjects may see, the handling and variables associated with several different species being tested under stress, and the reversed lighting conditions during testing. Birds tested in tunnels fly from a dark interior toward light, but free-flying wild birds fly from bright environmental light into darker habitats behind and reflected in the glass surface. Tunnel and field experiments mutually have confirmed the 2 x 4 Rule, and patterns created with elements separated by these dimensions should encourage glass manufacturers to prepare and offer bird-safe products for existing, remodeled, and new construction.

Bird of prey silhouettes placed on noise barriers along a highway into the city of Vienna (Wien), Austria.

Outside and inside view of Acopian BirdSavers applied to a residential home in Alburtis, Pennsylvania, USA.

Retrofit solutions include using decals as pattern elements, and a common practice is to use silhouettes of various shapes. One of the first was the use of birds of prey silhouettes. The use of decals, especially on Surface #1, where they would be most effective, is especially difficult to maintain because of outdoor conditions, such as wind and rain. Strips of tape are effective, relatively cheap, and sold as a preventive technique by the American Bird Conservancy (ABC, see collisions.abcbirds.org). Parachute cords hung vertically and separated by 10 cm (4 in) are an easily constructed and effective preventive method. Jeff Acopian, the first and most prominent user of parachute cords, promotes and describes how using these strings can be both aesthetically appealing and functionally effective. His Acopian BirdSavers website (www.birdsavers.com) accepts orders and instructs viewers on how to make their own. Hanging Mylar and cloth strips or monofilament line is similarly effective.

Outside and inside views of the external film CollidEscape applied on the Irma Broun-Kahn Education Center at Hawk Mountain Sanctuary, Kempton, Pennsylvania, USA.

Unique external film applied to Surface #1 is also an effective preventive measure. This one-way film was originally created for use in advertising by the 3M Company of Minneapolis, Minnesota. When viewed from outside, it can display any image. Viewed from the inside, it offers no obstruction and only a slight light reduction. These films appear regularly on buses advertising various commercial products and are then removed and replaced by images of another product. 3M does not sell this film for bird strike prevention specifically but will sell it to other companies for such a use. In the U.S., this film is sold to prevent bird strikes under the name CollidEscape; it can be ordered through Amazon. In Canada, the Convenience Group sells a similar product, branded as Feather Friendly. Both of these external films can be custom designed. Although 3M created the film for use over a period of weeks at most, its application depicting a pattern of white trees has been effectively preventing bird strikes at a row of windows at the Toronto Zoo for over twelve years. To my eye, the most unattractive application of this film is applying it with no pattern, giving the impression of a bright white sheet covering a window. Alternatively, any non-reflective matte or creative imaging can be used, just as when it is used to advertise a product, on the outside window surface. At the remodeled Irma Broun-Kahn Education Center on the grounds of the Hawk Mountain Sanctuary Association in Kempton, Pennsylvania, CollidEscape depicting a forest scene with flying raptors tastefully covering an extensive glass façade to protect birds.

Outside views of dot and line patterns of the external film prepared and applied by the Convenience Group (Subdivision: Feather Friendly) on commercial buildings in Toronto, Ontario, Canada.

Applying patterns to new panes for remodeling and new construction is an available method for architects to use in their building designs. The patterning can be prepared in any form and be completely effective in preventing strikes if the same 5–10 cm (2 x 4 in) spacing occurs between pattern elements. One type of patterning offers the human and bird eye a frosted glass appearance. Pattern elements are applied to new sheet glass by bonding ceramic paint to the glass surface under high temperature, resulting in what is called *ceramic frit*. With similar visual effect, frosted-like patterns also are applied on new sheet glass using acid etching. A number of glass manufacturers have used both these techniques to prepare sheet glass that is commercially available as bird-safe windows. A ceramic frit dot pattern effectively was used to prevent bird-window collisions in a new science building on the campus of Swarthmore College in Swarthmore, a municipality within greater Philadelphia.

Ceramic frit dot pattern on both sides of a corridor viewed from outside, and near and close inside on the campus of Muhlenberg College, Allentown, Pennsylvania, USA.

The ability of ultraviolet (UV) pattern elements to deter bird-window collisions continues to be investigated and discussed in the scientific literature. Skeptics about the use of UV to protect birds include the renowned ecologist Graham Martin from the University of Birmingham, U.K. In a 2011 review article in the journal *Ibis*, he described how a number of sensory ecology considerations suggest UV signals are not likely to be effective in alerting birds to the dangers of windows. Researchers Olle Hastad and Anders Odeen, from the Swedish University of Agricultural

Sciences and Uppsala University, respectively, writing in 2014 in the journal *PeerJ*, described the limitations of UV vision in birds and its ability to inform them about the presence and avoidance of windows. They emphasized how the visual system of birds in general consists of two discrete cone photoreceptors that are sensitive to UV, ultraviolet sensitive (UVS) and violet sensitive (VS). The maximum sensitivity for UVS photo pigments range from 355-380 nm, 402-420 nm for VS. Those species of birds with UVS vision include gulls, parrots, and passerines (perching birds). Over half of the approximately 10,500 species of birds are passerines. Most birds killed in window strikes are passerines. Most bird species have UV vision. The VS photoreceptors have a broad sensitivity that extends into the lower UV wavelengths as well. By attempting to interpret the complex components involved in determining what birds see, Hastad and Odeen judge that the mix of wavelengths perceived by the avian visual system result in low contrasts that likely translate into limited ability for birds to use UV signals as markers to avoid window collisions. They conclude that UV signals may be effective to deter bird-window collisions for UVS but not VS species.

The evidence and views of these respected scientists should not be ignored, and if nothing else they teach us how complex and sophisticated the avian visual system is compared with our own. Their work and insightful discussion underline how difficult it is to assess what birds actually see and further remind us of the obvious: We are not them and we never will be. Recall that bird anatomy, physiology, and behavior provide clues about what they can see and what they do with it. Although I am eager to learn from our sensory ecology and avian visual physiologist colleagues, none of these skeptics can explain why my experiments reveal that birds can in fact see and use UV signals to avoid windows.

A history of avian UV perception and its interpreted use in protecting birds from windows can assist in understanding its promise in creating short- and long-term prevention solutions. It was in the 1970s when scientists revealed that birds could see UV. I immediately recognized that this discovery might have special value in developing an elegant solution for bird-window collisions. Elegant because birds see UV patterning and humans do not. Especially germane to those early critics who were so concerned about interfering with the way people looked out their windows, if UV proved effective, humans could continue to enjoy all the benefits of visually unaltered sheet glass and still protect birds.

Another possibility that occurred to me at about this time was creating patterning that birds would see and we would not using nanoparticle technology. Nanoparticles are extremely small elements that theoretically could be used to develop a one-way window having visual markers on the outside but invisible to those looking out from the inside. I imagined material engineers being able to place nanoparticles into sheet glass in such a way that when windows were viewed from the outside, by birds and humans, a pattern would be visible. I speculated the pattern would result from interfering light waves, as occurs when we look through polarized sunglasses, or in the way that refracted and reflected wavelengths produce different colors or dark opaque areas, depending on the angle of view in the gorgets of hummingbirds or heads of Mallards. Looking for help to investigate this idea further, I searched the internet and found that Alfred University, just south of Rochester, New York, had researchers dedicated to the study of glass, nanoparticles, and the environment. That combination of interests seemed a perfect fit for my hopes that nanoparticles could play a role in bird-window collision prevention. But as with so many other outreach efforts, I seemed to have stimulated an initial interest, followed by no interest at all.

Looking further into using UV signals, I quickly learned that there was no way to test this option. From eyeglasses to sheet glass, manufacturers placed coatings, also referred to as glazing, on glass that absorbed UV to mitigate its harmful effects. They also created various other coatings that would save energy, the results being what's now known as low-E glass. It became obvious very early that a major problem was to find a product that reflected UV, because receptors in living organisms are stimulated by visual wavelengths of light. A window would have to reflect, not absorb, UV for birds, or any other creature to have the ability to see it. With only UV-absorbing coatings on conventional glass, I was stuck without a way to test my hypothetical "elegant solution."

Throughout the 1980s and '90s, I kept a hopeful open mind, but I reasoned that the prospects of UV actually working were remote. Ornithologists had found that reflected UV was used by various birds to identify foods such as berries, and by falcons to detect prey following mice urine trails. Female birds were known to compare the prowess of rival mates based in part on UV reflected off their feathers. At the other

Chapter 9

end of the visual electromagnetic spectrum, the wavelengths that were perceived as yellows, oranges, and reds have long been known by animal behaviorists to be used as warning colors. There is even a special term applied to these colors: *aposematic*. The combined warning pattern is referred to as *aposematic coloration*. With no way to test if UV could be used to create bird-safe glass, I answered questions about it by saying: "Not likely, given that the low end of the visual spectrum, the UVs, violets, and blues, have evolved to attract, while the upper end, the yellows, oranges, and reds, evolved to signal danger in the animal kingdom." This short general answer to an unanswerable question about the abilities of the diversity of life satisfied most questioners, myself included. A lengthy and more accurate reply, however, would need to include a far more complex explanation, describing how the physiology and behavioral ecology of color has evolved independently in many groups responding to varying selection pressures over vast amounts of time.

My inability to test UV ended one day in the early 2000s. During a morning coffee break, our geneticist, Irvin R. Schmoyer, asked me if I remembered how several years ago I helped him dig holes as part of a construction project to put an extension on his home. He explained that at that time he wanted to use plastic sheeting as skylights. No plastic sheeting, however, was capable of holding up to the harmful effects of UV. UV from the sun degrades plastic by turning it yellow and brittle to the point of cracking. As noted, glass was then, as now, coated to absorb UV. The absorption is not complete. Evidence that UV degrades plastic over time, even through glass coated with UV absorbers can be seen in the discolored and brittle plastic often seen on vehicle dashboards, and in faded curtains and carpeting in homes. Irv explained that if he were constructing the room now, he could use plastic sheeting, because protective UV-absorbing coatings were now being used on that material. This was big news, very big news, for me. I immediately began thinking about how to use the new UV-protected plastic sheeting to test the ability of UV to prevent bird strikes.

I knew from film and glass manufacturers that conventional sheet glass reflects about 13% of UV. I reasoned that I could use the UV-absorbing plastic to create a pattern of UV-reflecting and UV-absorbing areas across the surface of a conventional window. The strength of UV reflection would be limited to 13% and be adjacent to the UV-absorbing plastic that would reflect little or no UV. I asked our carpentry shop at the college to obtain the thinnest plastic sheeting possible and cut it into 2.5 cm (1 in) strips. I applied the UV-absorbing strips over conventional panes separated by 5 cm (2 in). The resulting UV pattern was 5 cm (2 in) UV-reflecting stripe alternating with 2.5 cm (1 in) UV-absorbing stripe, oriented in vertical columns across the entire single pane. I tested these modified panes in a field experiment. The results revealed mixed effectiveness, and I attributed the outcome to a relatively low level of UV-reflectance. Nonetheless, the results were encouraging enough to suggest that UV may actually work. To investigate further I needed a new and ideally improved UV signal to test.

In yet another good surprise, which I shared earlier, I received unexpected help in the form of a call from an external film company in Martinsville, Virginia. The caller introduced himself as Tony Port, head of research and development at CPFilms. He said he'd heard me interviewed by NPR science reporter John Nielsen and described how he was keenly aware of bird-window strikes. He enjoyed a personal interest in birds from his days growing up in England. He wondered whether CPFilms already had a product that could make glass safe for birds. I accepted Tony's invitation to visit Martinsville, where he and his colleagues described how they had films that completely absorbed UV similar to sheet glass coatings. Based on my previous experiments with visual patterns, I predicted that without a differentiating pattern, a complete absorption of UV would be ineffective. With only suspicions, I was willing to try. At the time, there were no other options when it came to evaluating UV as a protective method.

As predicted, the completely UV absorbing film did not work. Then Tony surprised me again by preparing a number of prototypes with different UV signals. Among them were films that reflected UV uniformly over the entire window. Another pattern we called S-1R consisted of 2.5 cm (1 in) UV-reflecting vertical stripes separated by 5.0 cm (2 in) complete UV-absorbing stripes. Another similar pattern we labeled S-2R that consisted of 5 cm (2 in) UV-reflecting stripes separated by 2.5 cm (1 in) UV-absorbing stripes. Still another pattern formed a grid with 2.5 cm (1 in) UV-absorbing stripes crisscrossing over the window and highlighting UV-reflecting blocks 10 cm (4 in) across and 8 cm (3 in) down. CPFilms reported their UV measuring instrument, a spectrophotometer, recorded a strength of 80% UV-reflection over the

300-400 nanometers (nm) range and near to complete UV-absorption for the adjacent areas.

In 2009, I described the UV details in a paper in the *Wilson Journal of Ornithology*. While conducting subsequent experiments, I asked CPFilms to remeasure their films, they discovered an error. The reflected strength of UV-reflection over the 300-400 nm range was in fact 20%–40% instead of 80%. Nonetheless, the 20–40% UV-reflection adjacent to UV absorption produced a UV pattern that effectively deterred bird strikes. The level of success was measured by comparing the number of strikes at the UV test pane compared with those at clear and/or reflective control panes. I repeated the experiment and confirmed that, compared with an unaltered clear control, the S-1R film significantly reduced the risk of bird strikes by 97% in the first experiment and 92% in the second. The S-2R film significantly reduced the risk of bird strikes by 98% and 92%, respectively. The grid tested in the first experiment reduced the risk of strikes compared to the control by 93%. It was both remarkably surprising and clear to me that patterns visible to birds and not humans, created with reflecting and absorbing UV elements, were as effective in deterring bird-window collisions as those patterns visible to both birds and humans. Looking at these prototype films, they appeared invisible, as if the window was unmarked. I was elated. I seem to have found the elegant solution, even though I was at first highly skeptical. Mostly, I was grateful for not dismissing the possibility.

UV-reflecting and UV-absorbing pattern in direct sunlight and shade of external film prototype (S-1R) prepared by CPFilms, Eastman Chemical, Martinsville, Virginia, USA, photographed by Jeffrey Manning, SurfaceCareUSA.

Solutions – Solving the Problem

After learning about the results, Tony was as thrilled as I was and recommended that I seek a patent documenting my discovery of a UV pattern that prevented bird-window collisions. He justified my committing the effort to a patent by explaining that even though CPFilms committed to including me in any revenue that would come from what they called the "bird film," a legal document would reinforce their fulfillment of this promise. The prospect of receiving a financial return on the incalculable personal expenses I have incurred over the years studying this issue was a welcome hope. At this point, however, just acquiring a patent seemed formidable and highly improbable.

As advised, I began investigating patents. I learned from my reading and from asking friends at my workplace who had experience with patents that on average it was common for corporations to invest $60,000 or more in acquiring a patent. I also quickly learned that the U.S. has a reputation for being a litigious society. Corporations that obtain patents for their products characteristically have large legal teams divided into two camps, one to fiercely protect the company's own patents, the other to finding ways to reduce costs by not honoring the patents of others that might be associated with their products. Then there was the need to defend a patent from those who ignore the legal rights it bestowed. Defending a patent means you have to sue an offender. To do so, you need enough financial resources to battle in court. I had no illusions that I had the means to defend a patent. If I pursued one, it would have to be with the hope that the firm I partnered with would honor it. Tony and CPFilms offered me that hope.

I began looking into patents. At first, I reasoned that this is the U.S. Any citizen is supposed to be able to do anything. After all, did not Abraham Lincoln practice law without a formal degree from a law school? In today's world, knowledge is regularly said to be just "a Google away." Surely, I could apply for a patent on my own without the help of a legal professional and the accompanying substantial fees.

I tried and failed, even with the guidance I received from a friendly and informative U.S. patent examiner. Patent examiners are responsible for investigating the claim that what you apply to patent is new and warranted. It did not take me long to realize the task was overwhelming. The basic legal language and forms alone were daunting. My confidence in acquiring a patent was barely a flicker, but not completely extinguished.

Chapter 9

The success of the CPFilms prototype provided documented evidence that UV patterns could be used to prevent bird-window strikes. While investigating patents, I learned about a window product using UV called Ornilux, manufactured by a German company, Arnold Glas. Through an exchange about bird-window collisions with a New York City architectural firm, Fox and Fowle, I learned that Ornilux had been installed with reported success at a building at the Bronx Zoo. I asked Fox and Fowle to help me contact Arnold Glas in hopes that I could evaluate their special UV panes. But I had no success in learning more about Ornilux or contacting Arnold Glas for over a year.

Then, another good surprise: a representative of Arnold Glas contacted me. They suggested to meet at Muhlenberg College. I was excited, overly delighted and readily accepted. The owner, Hans-Joachim Arnold, and his principal officers came to us, and the result of the meeting was an agreement that I would test their bird-safe glass using our field experiment design. They explained that they had modified it from what had been installed at the Bronx Zoo, from a UV striped pattern to a Mikado, which they described as having a visual pattern similar to what scattered slim sticks, pickup sticks, would look like if you could see the pattern in the UV. I had some concerns, because the coating that produced the UV signal was not on Surface #1. It was applied to an inner surface, which meant that the promising signal could be masked by reflections off Surface #1. As noted, any pattern applied to any inner surface of multipane windows runs the risk of being of little or no value in alerting birds to collision danger. Even with this concern, I expected my field experiments would further confirm their claims, which had already produced positive tunnel testing results at the world famous Max Planck Institute for Ornithology.

That meeting with Arnold Glas at Muhlenberg College was memorable, partly because of our mutual interest in saving birds from windows, but also from a more light-hearted experience. The day we met, July 2, 2010, was a swelteringly hot day in Allentown. Our meeting room, however, was remarkably cold. So cold in fact that our guests were visibly shivering. My call home resulted in Renee bringing over several sweaters to rescue our frozen guests.

Mr. Arnold explained that he obtained and applied ideas about using UV to prevent bird-window strikes from an attorney friend, Alfred Meyerhuber, who in turn had been inspired by my research that had considered UV as a means to preventing strikes. Meyerhuber judged the use of UV to be original enough to warrant a patent, which he applied for and received. I explained that I, too, was attempting to acquire a patent using UV, but so far the U.S. Patent Examiner had turned me down. The examiner justified rejecting my application because Meyerhuber of Germany had already patented the UV idea.

Another exchange at the meeting was relevant to the bird-safe claims of their Ornilux. I vividly recall when Mr. Arnold and his staff became alarmed after I, perhaps too passionately, tried to explain the problems that might occur regarding the effectiveness of Ornilux Mikado applied to the inner surface of a window complex. I have found that most glass manufacturers prefer modifying the inner surfaces of windows, to protect them from weathering. I was mortified and embarrassed to learn that they thought I was angry, based on the manner of my enthusiastic explanation. Mercifully, with capable translators I was able to assure them that I would be the very last person on Earth to offend anyone intentionally. We easily overcame the misunderstanding, and they promised to send me a pane of Ornilux Mikado to evaluate in our field test.

The results of the field tests provided yet another surprise, and this time not good. Not only was the Ornilux Mikado pane not bird-safe, it actually posed a 28% greater risk than clear see-through conventional windows. In a follow-up experiment, Ornilux Mikado did much better when it was placed over dark panels to simulate covering a room with a dark interior. Installed in this way, Ornilux Mikado reflected the facing habitat and sky and reduced the risk of a bird strike by 55% compared to a clear control window. Our field test results were very different than what ABC had documented by tunnel testing. ABC's tunnel test found a see-through Ornilux Mikado reduced the risk of a strike by 58% and 66%. These results were markedly different, and opposite, in fact, from our see-through findings, but similar to what we found when Ornilux Mikado was a reflective pane covering a dark interior space. When I wrote our results up for publication that appeared in the 2013 *Wilson Journal of Ornithology*, co-authored with my research colleague Peter G. Saenger, we acknowledged not knowing why Ornilux Mikado was more bird-safe when installed as a reflective window covering a dark room compared to see-through panes. Perhaps the UV pattern offered some visual cues to birds with the dark background. The comparative results evaluating

Chapter 9

habitat seen behind Ornilux Mikado windows from ABC tunnel testing and our field experiments were not just radically different, they were alarming because the claim was made that the pane was bird-safe when it was not. In our paper, we used the difference in results from these two experimental designs to highlight why tunnel testing should not be used as a final assessment. As useful as tunnel testing is for evaluating several methods in a timely manner, its value is in the preliminary assessment of many options. Testing sheet glass in tunnels cannot accurately simulate what happens with windows installed in human structures. A field experiment is required as a final assessment, to be confident a preventive method works. Of even more concern is the fact that some stakeholders, such as ABC and their architecture and building industry supporters, advocate using tunnel testing as an industry standard, to evaluate the level of safety offered by alternative bird-safe windows. To repeat, the risk of using tunnel testing for this purpose is to identify a product as bird-safe when in fact it is not, such as Ornilux Mikado was in 2010 when tested in our field experiment.

So why was Ornilux Mikado not bird-safe? I looked for the answers in the details of the UV signal it projected to birds, in hopes of comparing them to what the successful CPFilm prototype offered. A consequence of this investigation was that I lost my ability to work with the Arnold Glas company, a professional and personal tragedy for me. I am still very saddened and regret this outcome. I very much liked and respected Mr. Arnold and everyone I interacted with at this company. I believe, and hope, that our inability to work together resulted from another misunderstanding I was not able to overcome. Perhaps what best explains our estrangement was not having the opportunity to work out our differences in person or not having the services of an accurate and comprehending translator.

Our parting resulted from my asking for the spectral details of the UV signal that Ornilux Mikado reflected. To obtain this information, Mr. Arnold's representatives required that I sign a non-disclosure statement, which I had done for others and expected to do with them. I was cautious in all our deliberations, asking that Arnold Glas agree to select freedoms regarding what I could and could not share with others. I was sensitive to having this freedom addressed and agreed upon, because other researchers had informed me that a signed non-disclosure agreement with Arnold

Glas prohibited them from publishing their findings in a peer-reviewed scientific journal. Consequently, I wanted to be sure that what I learned about the effectiveness of UV signals protecting birds from windows could be shared with others for the universal good of the birds we were trying to save. As with agreements I signed with other companies, I had no conflict agreeing to keep all manufacturing or other details confidential. But I firmly believed it was my obligation to share why, if I could determine it, the UV signal from Ornilux Mikado did not work. At least in the case of see-through panels, why was it more deadly than conventional panes? I wanted to avoid signing an agreement that would stop me from disclosing what an effective UV signal had to be. My rationale to Mr. Arnold and his representatives was that I had spent a great deal of my professional career, arguably a lifetime, trying to describe and prevent bird-window collisions. I did not want to compromise my responsibility to share what I learned with the scientific community that I identified. My training directed how research must be part of the public domain, to be available to Arnold Glas and any others preparing effective products to save birds from windows. I explained that what I hoped to learn from them about their UV signal would offer the opportunity to guide them in modifying Ornilux Mikado to be more effective. None of my efforts worked, and to my crushing disappointment and regret, they severed their relationship with me.

With no information forthcoming from Arnold Glas, the question of why Ornilux Mikado was not bird-safe glass remained a mystery. In an exchange with Michael Mesure at FLAP, I learned that Arnold Glas freely gave samples of Ornilux Mikado to those they judged would benefit from learning about their product. They gave samples to architects and to FLAP as an organization that could recommend their bird-safe product. Michael informed me that he had a small sample. He was willing to let me borrow it to measure the UV details I sought. I received the sample, and promptly had it examined and measured by a colleague in our Department of Chemistry at my work and an independent laboratory at which I had received spectrophotometer readings for the CPFilms UV external film prototype. The spectral readings were remarkably revealing. The UV signal from Ornilux Mikado reflected a maximum 7%–22% UV from 300-400 nm, reaching above the strength of 20% reflection only at 397 nm. By contrast, the prototypes from CPFilms had a UV-reflecting component of 20%–40% over 300-400 nm. These differing details were documented in our 2013 *Wilson Journal of Ornithology* article. Our interpretation described how the inability of Ornilux Mikado installed in a see-through setting to deter bird strikes is due to a lower level of reflected UV that is available for bird perception. Additionally, we judged that to offer an effective collision deterrence, the UV-reflecting component of the signal minimally must be 20–40%, and be adjacent to contrasting areas of UV-absorption to further highlight the UV signal overall.

It was just after my testing of Ornilux Mikado that I discovered a note in my patent papers identifying a patent attorney, one James C. Simmons. The note described him as a career patent attorney at a locally prominent internationally distinguished corporation, Air Products and Chemicals. He had retired from Air Products, but was still practicing law at an Allentown firm, Design IP. My contact information had his telephone number, and since the U.S. Patent Office had all but extinguished my hopes, he was my last option to keep what now became my patent dream alive. He heard me out and agreed to meet to discuss what, if anything, could be done. I explained that my patent application had been rejected, as best I could tell, because a German patent had precedent and my application did not have any novel claims to make my submission worthy of recognition as a new invention. Mr. Simmons agreed with enough of what I shared to accept me as a client. Through my description, he came to see my application had merit based on the critically important need to create a UV pattern with specific dimensions of UV-reflection and UV-absorption. These critical dimensions were revealed from my experimenting with the CPFilms prototype and Ornilux Mikado. Where the German patent claimed to use UV patterning to deter bird-window collisions, it did not identify the need to have a specific UV pattern to be effective. I argued that the strength and spacing between UV pattern elements was crucial and without both, a UV signal would not work.

Mr. Simmons drafted and submitted a modified U.S. patent application in my name. More than a few months later, he called to tell me he had received a rejection from the patent office. He recommended we meet again. Prior to the meeting, I carefully read over the patent examiner's reasons for rejection. At least I felt the examiner did not understand the importance and uniqueness of having a specific UV signal to be effective. I suggested to Mr. Simmons that we call and talk to the examiner in hopes of providing more of an explanation that might not have been clear in our written application. It seemed clear to me that somehow the examiner did not understand the necessity of UV pattern spacing to be effective. Consequently, his reasoning for rejection using, Dr. Meyerhuber's patent were based on generalities rather than specific details. I recommended making a call and talking directly to the examiner, based on the previously cordial and helpful exchange that I had had with the patent examiner that rejected my first go-it-alone application. Mr. Simmons very clearly and forcefully replied that he would not do that. He rather sternly described how in the decades that he had successfully practiced patent law, he never talked directly with a patent examiner. All communications were formally in writing. OK, what did I know? My final offering on this subject was that I figured it couldn't hurt. After all, to date I had only been turned down: 2 for them, 0 for me. Neither of us had an idea about what to do next, except of course to give up. We parted, agreeing independently to consider what, if any, other action was warranted.

Just a day had passed when I received a call from Mr. Simmons. He offered no reason for a change of mind, but simply said he had made arrangements to talk to the patent examiner by conference call from his office. He asked me to be there as our next step. We made the call. Mr. Simmons introduced the issue of criticality of UV signal dimensions that we were arguing made my application unique, and then asked me to explain the details and rationale to the examiner. I did, and to the surprise of both of us, within a ten-minute exchange the examiner agreed and informed us the patent would be awarded. This original and a subsequent modifying clarification of the patent has value because it specifically documents what the quantifying details of the UV signal must be to deter bird strikes.

Making the UV solution available for commercial sale to protect birds has not been as successful. CPFilms and the firms that have purchased the company over the years—currently it is a division of Eastman Chemical Company—have not manufactured or offered their successful bird film prototype for sale. My supporter within the firm, Tony Port, has had a distinguished career with the company and was promoted a few years ago and moved to Palo Alto, California. He continues to advocate for their "bird film," and over time, has periodically called me when a new manager was appointed with influence over which new products they would offer. His calls to me are always hopeful because he makes them when he believes the newly appointed manager is interested in and sympathetic to the bird film. He certainly is, and these calls give him the chance to reassure me that he is still on the case promoting its manufacture and sale.

So what keeps CPFilms from making and selling their bird film? I have asked every prominent conservation professional I know to contact and support production of the film. These include principal administrators and scientists of our most prestigious conservation organizations, and U.S. government agencies, the National Audubon Society and the U.S. Fish and Wildlife Service are just two. Additionally, with every appropriate opportunity, I write CPFilms to stimulate their interest in bird film production. These opportunities arise when I publish data emphasizing the effectiveness of their product or in a review article that invariably includes encouragement to save billions of birds using their bird film or other products having a similarly effective UV signal. None of these efforts have moved CPFilms. Tony tells me that the bird film has not been discarded but remains on the shelf for a potential future commitment to manufacture. What information I have received from him and other principals at CPFilms is that their lack of commitment and action is a result of being unable to determine what return they would get from an investment in a product attractive to the "bunny and tree huggers" of the world, a constituency they are insufficiently familiar with. This lack of knowledge is an incalculable hole in their business planning. I am perplexed and frustrated by their lack of comprehension of an unlimited potential market for their external film. The existing windows of the world need to be retrofitted to protect birds. Even if only those aware of the issue are considered, it is difficult to understand why the film would not have wide use on government and private buildings, in North America and around the world.

Perhaps we will soon have another good surprise, learning that growing awareness and commitment to saving birds from windows has encouraged Eastman Chemical, or whoever else might acquire the CPFilms division, to manufacture an effective external film using UV signals. According to Tony, decisions about what to manufacture are made in March each year, and only one or two alternatives are chosen. An entire plant is devoted to making one product at a time. I am still hopeful that the bird film will eventually make the cut, but Tony recently told me he doesn't expect Eastman Chemical to ever produce the bird film unless some competitor encourages them to reconsider. Other manufacturers have the technology to do so, however. Here's hoping that happens to curb the loss of bird life at existing killer buildings.

If an effective external film eventually is made available, it will still be only a short-term solution. As with all external films, it will need periodic replacement, unlike a novel sheet glass that had effective UV patterns applied to Surface #1. When and if these panes are made available, they will offer a long-term solution for use in remodeled and new construction. A few glass manufacturers have shown interest and have been researching this promising long-term solution. As with external films, however, as I write now in July 2019, no effective new panes using UV patterns were available for sale to protect birds, with one exception. That is a product called AviProtek T, manufactured and sold as bird-safe windows by Walker Glass Company of Montreal, Quebec, Canada. In prototype form, these panes reduce the risk of bird strikes by 62% to 80%, calculated from comparing the number of strikes at the UV pane with that of an unaltered control pane. Although other manufacturers have similar windows in development, AviProtek T is the only bird-safe window using a UV component that is available for remodeling and new construction. The hope is that Walker or other glass manufacturers will create windows using UV signals that will be even more effective in preventing bird-window collisions.

Clearly, to any reasonable person with a fundamental knowledge of the issue, there is hope that enough interest among enough citizens of the world will encourage the production of these bird-safe products. The critical mass of support is just not there yet. Inciting meaningful action will require effective education using every available means, from media to celebrity support. Like most, I've had little prospect of being able to convince a celebrity that the bird-glass issue is worth their time. An exception occurred a few years ago when my employer, Muhlenberg College, awarded former Senator Bill Bradley an honorary degree at our 2016 commencement. I was part of the platform party. Senator Bradley and I stood next to each other during robing, and I struggled with bringing up the bird-glass issue with him, reluctant to intrude on his personal time and sensitive to in any way detracting from the joy of his award experience with us. I suppressed that urge, reasoning that this was a rare opportunity to attract the help of someone who could theoretically save far more bird lives than I was likely to in a lifetime. After introducing myself, I congratulated him on his worthy achievements that fostered the honorary doctorate and told him I was aware of his interest in environmental issues. I asked if he would be interested in learning more about an environmental issue that needed help from him or another prominent person. His silence permitted me to ask if he had ever heard about birds being killed flying into windows. I invited him to look into it and contact me if he developed any interest. No followup from Mr. Bradley. To date, I've had no other opportunity to convince a celebrity with known environmental interests who might enthusiastically offer to save more bird lives from windows.

Perhaps yet another good surprise awaits. Capturing the meaningful attention of an influential person certainly would save far more bird lives from windows than we ordinary citizens can hope to do. Aided by a celebrity or not, I have described how there are growing solutions to this problem. Unless we can identify and employ a completely effective bird-safe solution everywhere such as the "elegant solution" represented by UV-reflecting products, it is likely that humans will have to sacrifice some aesthetic appearance to their dwellings. Otherwise, birds will continue to have to sacrifice their lives at untreated sheet glass.

Practical Solutions

Windows of any size are lethal hazards for free-flying birds. Existing windows require retrofitting to make them safe. Remodeling and new construction require the use of bird-safe glass to make them safe. Windows offer a see-through effect, where habitat and sky are visible behind clear panes, and patterns can be placed on any inside or outside surface. Where windows reflect the facing habitat and sky, patterns must be placed on the outside surface (Surface #1). See-through conditions occur at outside railings, noise barriers along highways and rail lines, corridors (linkways) between buildings, atria, glass-walled interior rooms, and where clear panes meet in corners. Most window installations result in a reflection off of Surface #1 when the panes cover a dark interior. Even a perfectly clear window will act like a mirror when viewed from outside if covering a dark interior space. Patterns that effectively deter bird strikes must uniformly cover the surface of a window, and the elements making up the pattern must follow the 2 x 4 Rule. The 2 x 4 Rule states that elements making up a pattern must be separated by 2 inches (5 cm) if oriented in horizontal rows, and by 4 inches (10 cm) if oriented in vertical columns. Reducing the spacing between vertical pattern elements to that of horizontal elements results in modifying the 2 x 4 Rule to the 2 x 2 Rule. Reduced vertical spacing is expected to add more protection for hummingbirds and other hummingbird-size species. Effective deterrent patterns can be applied to existing windows with soap, tempera paint, tape, films, and cords, strings, or ribbons. There are currently no commercially available effective retrofit deterrents that use UV patterns consisting of elements visible to birds but not humans. New glass used for remodeling and new construction have pattern elements applied as ceramic frit, acid etching, and UV-reflecting and UV-absorbing coatings. Several glass manufacturers offer effective visual patterns. As noted, as I write, only Walker Glass's AviProtek T windows effectively use UV patterns to deter bird-window strikes. Several other manufacturers have UV patterned glass in development. Whether retrofitting or using novel bird-safe glass in remodeling and new construction, minimally apply deterrent patterns to the first 16 meters (52.5 ft), or the first four to five stories associated with surrounding mature vegetation. A sampling of bird-window collision deterrent methods and their relative costs is printed and internet referenced in the Resources section at the back of this book.

Chapter 10

Legal Protection of the Vulnerable and Defenseless

History has revealed that the use of the legal system is a far more powerful means of stimulating action to protect birds from windows than relying on the voluntary efforts of the many constituencies involved in this important conservation issue. Current legislation to protect birds from the threat of injury and death from striking windows is available at the federal, regional (provincial, state), and local municipal levels. Complexity and confusion have rendered national and regional legislation questionable and to date mostly ineffective, but mandatory ordinances and zoning regulations are being adopted by a growing number of municipalities in North America, and these guiding rules are serving as models for others elsewhere around the world.

At the federal level, international treaties and laws protect most species of wild birds in North America except for select introduced and escaped exotic species. In the U.S., birds receive protection under the authority of the 1918 Migratory Bird Treaty Act (MBTA). This iconic legislation originally was crafted to protect birds from the millinery trade when bird feathers, such as the elaborate plumes of egrets, were literally worth more than their weight in gold. In 1973, the U.S. Endangered Species Act (ESA) was passed to protect those birds whose populations were threatened with extinction. Both these legal measures were amended over time and remain eligible for further changes or elimination.

The equivalent to the MBTA in Canada is the Migratory Bird Conservation Act (MBCA). Japan, Mexico, and Russia have their own respective equivalents as signatories to this international treaty. Other applicable Canadian federal statutes protecting birds similar to the ESA are the Environmental Protection Act (EPA) and the Species at Risk Act (SARA). In the 1970s, member states of the European Union

(EU) adopted The Bird Directive to protect birds as a shared heritage. Among other purposes, collectively these legislative agreements emphasize how avian conservation requires international cooperation. Multi-country collaboration is essential for animals such as birds that do not recognize and cross several sovereign borders during their annual life cycle.

At present, there is controversy in Canada and the U.S. about whether, individually or collectively, these federal legislative documents address the unintentional as well as the intentional killing of protected birds. The formal designation for killing is *take*, more specifically *incidental take*. In 1999 and 2015, Larry Martin Corcoran and Devin T. Kenney writing in the *Denver University Law Review* and *Journal of Animal and Natural Resource Law*, respectively, addressed how different legal professionals and federal courts have dealt with assigning accountability for killing protected birds. Interpreting these statutes broadly means holding individuals, governments, and companies criminally accountable when birds die, even if their actions carried no intent to kill. Interpreting the statutes narrowly means defining only the direct killing of migratory birds as criminal behavior, without accountability or consequences for unintentional killing.

In the U.S., 2018 marked the 100th anniversary of the Migratory Bird Treaty Act (MBTA), a federal law in the United States that implements a 1916 treaty with Great Britain. The short preamble to the treaty says:

> Whereas, many species of birds in the course of their annual migrations traverse certain parts of the Dominion of Canada and the United States; and
>
> Whereas, many of these species are of great value as a source of food or in destroying insects which are injurious to forests and forage plants on the public domain, as well as to agricultural crops, in both Canada and the United States, but are nevertheless in danger of extermination through lack of adequate protection during the nesting season or while on their way to and from their breeding grounds;
>
> His Majesty the King of the United Kingdom of Great Britain and Ireland and of the British dominions beyond the seas, Emperor of India, and the United States of America, being desirous of saving from indiscriminate slaughter and of insuring the preservation of such migratory birds as are

Chapter 10

either useful to man or are harmless, have resolved to adopt some uniform system of protection which shall effectively accomplish such objects.

The treaty was a device used by proponents of the law to safeguard it from constitutional challenge. An earlier law, known as Weeks-McLean, was ruled by two federal courts to be unconstitutional. In the midst of World War I, Senator Elihu Root (later Secretary of State) suggested a treaty, giving the federal government the constitutional authority needed to act under the treaty power. After negotiation with Great Britain, the treaty was signed on August 16, 1916, and the Senate consented to ratification only 13 days later; it entered into force on December 7, 1916. Congress would fulfill the legal obligation to enact implementing legislation in June 1918. The Biological Survey, a bureau of the Department of Agriculture, issued the very first regulations on October 23, 1918. The constitutionality of the law was fortified by the Supreme Court in 1920 in the case of *Missouri v. Hollander*.

The new law expressly prohibited anyone to "hunt, take, capture, kill, attempt to take, capture, or kill ... any migratory bird, included in the terms of the Convention between the United States and Great Britain for the protection of migratory birds ...". Congress, however, did not think to define the term "take." The implementing regulations promulgated by the Secretary of Agriculture actually defined "take" in an utterly tautological manner: "Take—the pursuit, hunting, capture, or killing of migratory birds in the manner and by the means specifically permitted." In other words, take meant nothing more than one of the other four specified acts.

The MBTA has been amended numerous times, including additions that incorporate subsequent treaties with Japan, Mexico, and the former Soviet Union. It has been amended to exclude non-native species. Penalty provisions have been amended. It has been amended to provide that the strict liability provisions apply only to misdemeanors; a felony conviction requires proof of a conscious violation. Through the many amendments, Congress has never seen fit to define the term "take." Another federal regulation defines the word "take" as "pursue, hunt, shoot, wound, kill, trap, capture, or collect, or attempt to pursue, hunt, shoot, wound, kill, trap, capture, or collect." Again, *take* seems to have no meaning apart from the other activities.

Surely the word "kill" means "take a life?" If a homeowner fells a tree and a nest of baby birds falls to the ground, causing the young birds to die, then the homeowner has killed those birds. As to misdemeanor violations, the MBTA is a "strict liability" statute, so no intent to kill is necessary to prove the violation. The homeowner is liable even if he did not know the nest was in the tree. And if a homeowner places glass windows in her house, and a bird flies into the window and is killed, the homeowner is legally responsible. Birds are electrocuted on power poles. The power companies are legally responsible. Birds are killed flying into telecommunications towers and wind turbines. The owners of those structures are legally responsible.

Maybe not. These are all cases of otherwise legal activities that have the unintended consequence of killing birds. The term to describe this kind of killing is "incidental take." Because the statute and the current regulations say nothing one way or the other about *take* incidental to otherwise lawful activities, no one knows if such activities are, in fact, violations of the law.

Then again, maybe not. For decades, the USFWS used this uncertainty as leverage to convince industries to enter into discussions with the agencies about developing practices to minimize incidental take. The agency was able to persuade some industries to develop, test, and implement bird-safe practices in exchange for an implied promise that so long as they used those practices, the agency would use its prosecutorial discretion and refrain from prosecuting for bird deaths. As a result, long-line fisheries developed scaring devices to keep diving seabirds away from the miles of hooks set for fish. Electric transmission line companies developed best practices, such as barriers to prevent the wings of large birds from touching the conductors. Other industries were less amenable to the process. The telecommunications tower industry met with the agency but would not agree to investigating or making changes to their facilities. The wind turbine industry participated in extensive discussions, facilitated by an outside organization funded in large part by the industry, but committed to no bird-safe standards or changes. The USFWS chose to prosecute only when companies failed to implement these industry bird-safe practices or when industries that had failed to participate in this process caused avian mortality.

The USFWS did not address the window glass problem, probably because it involved millions upon millions of built structures owned by millions upon millions of companies and individuals. Discussions pertaining to window glass, primarily with the American Institute of Architects (AIA) and window glass manufacturers, were initiated by scientists and later by bird conservation organizations. It is highly doubtful that the USFWS would ever have prosecuted bird deaths caused by window glass, except perhaps in some highly publicized cases of glass structures such as sports stadiums, and then only if the agency had first attempted to persuade the building owners to use bird-safe materials and practices. Even if there had been a coordinated effort by the USFWS to address the window glass problem, it is unlikely that the agency would ever have pursued action against homeowners. Through 2016, the efforts made by the USFWS were limited to an informational pamphlet called *Reducing Bird Collisions with Buildings and Building Glass: Best Practices* and some information on the agency's website. Even when the Minnesota Vikings planned a shiny glass stadium that was predicted to kill birds extensively–and it has–the USFWS filed no formal objections.

The MBTA permitted the USFWS to give potentially violating industries an incentive to develop and implement bird-safe practices in exchange for the assurance of no prosecution for avian mortality. Minimally, this policy held potential for reducing avian mortality from human activities, even if some industries were reluctant or unwilling to accept it.

Maybe not. Over time, the USFWS did prosecute some companies that had failed to implement their industry's bird-safe practices or that had not participated in that negotiated approach at all. These cases included a power line company, companies whose chemical waste pits attracted birds, and wind turbine companies. Not surprisingly, some of those companies took the stance that the MBTA did not apply to incidental take, and some of the courts that considered the question agreed. In the 1998 case of United States versus Moon Lake Electric Association, Inc., Judge Babcock interpreted the MBTA broadly and decided the power-line electrocution death of migratory birds warranted strict criminal liability. A contrary judgment by Judge Hovland, U.S. District Judge for the District of North Dakota, ruled in favor of several gas and oil firms when a suit using the MBTA was brought against them for the unintentional killing of birds at contaminated water pits (United States versus Brigham Oil and Gas, 2013). In this case Judge Hovland interpreted the MBTA narrowly, ruling the MBTA is only able to address the direct, not indirect or unintentional, killing of migratory birds as criminal behavior.

Overall, the issue has been considered by several federal courts of appeal. Some have judged that the MBTA applies to unintentional bird deaths that result from otherwise lawful activity. Others have judged that the law applies only to actions taken for the purposeful killing of birds. Under federal law, the decisions of circuit courts apply only in the geographical boundaries of the circuit. Thus, as of 2019, the MBTA applies to incidental take only in a few states, among them Connecticut, New York, Vermont, Colorado, Kansas, New Mexico, Oklahoma, Utah, and Wyoming.

Then again, maybe not. As the Obama Administration was ending in 2017, the USFWS issued a policy document known as an M-Opinion that stated that incidental take is prohibited under the MBTA. An M-Opinion is internal guidance to the department agencies. It does not have the force of law. In essence, it is legal advice from the lawyers who work for the Department of the Interior (DOI), telling divisions that the department will support this interpretation; it does not require courts to accept it. That M-Opinion (also since known as that "Tompkins Opinion," named after the DOI's chief lawyer) noted that surely "take" must have meant something other than the listed activities, because the law also pertained to songbirds, which are not hunted or otherwise deliberately taken. This policy asserted that, provided there is a sufficient causal connection, i.e., that the act caused the bird's death, the MBTA prohibits incidental take of protected species of birds.

Then yet again, maybe not. For the purpose of review, the Trump administration withdrew that M-Opinion on February 6, 2017, and then it has not been reinstated. On December 22, 2017, the DOI issued a new M-Opinion (also since known as the "Jorjani Opinion," named after the then current DOI chief lawyer) stating unequivocally that the MBTA applies only to affirmative actions (as opposed to inaction or neglect) that have as their purpose the taking or killing of migratory birds. The then current government concluded that incidental take is not covered under the MBTA. In opposition to this decision, several conservation organizations (National Audubon Society, American Bird Conservancy,

Center for Biological Diversity, Defenders of Wildlife) and U.S. states (California, Illinois, Massachusetts, New Jersey, New Mexico, New York, Oregon) have filed suit. The case remained pending as of August 2018. Whatever the outcome of this lawsuit, unless enough U.S. citizens convince the Administration, lawmakers, and government professionals charged with interpreting and enforcing legislation, it is highly unlikely the incidental take of protected birds will be prosecuted under the MBTA.

Consequently, until changes occur, the MBTA does not afford birds protection from incidental take. Until changes become possible, the pending lawsuit could reach the Supreme Court. If the Supreme Court agrees to take the case and judges no liability for incidental take, only an amendment of the statute by Congress could create the legal foundation to identify incidental take of species protected under the MBTA.

The legal case for federal prosecution of window-kills is stronger if the victim is one of the 101 species protected under the Endangered Species Act (ESA). Although the ESA does not specifically prohibit incidental take, Section 10 requires USFWS permits for any unintentional kills of ESA-listed species while conducting otherwise lawful activity. Certainly, building a structure with windows is a lawful activity. In practice, however, the USFWS requires permits for the incidental take of ESA-listed species only in conjunction with select large-scale or land development projects where the species is known to occur. For example, some wind energy projects have applied for and received incidental take permits to erect wind turbines in places where endangered bats are present. Most wind energy projects, however, have not applied for and do not have USFWS permits. Guidance issued by the USFWS in 2018 to agency staff stated that they were to proactively advise potential applicants that a permit to cover incidental take is only appropriate when an activity is likely to result in the killing of ESA-listed wildlife, and it is the potential applicant's decision whether to apply for a permit to cover incidental take. The expectation and hope is that a company that wants to avoid prosecution will seek scientific expertise to evaluate the likelihood of killing ESA-listed wildlife, and if so seek a permit. The low numbers of incidental take permit applications suggest that companies engaged in projects where ESA-listed species are at lethal risk either do not assess the potential for take, or they know it is highly unlikely that the take will be detected. Obviously, without detection there would be no grounds for prosecution. Although ESA-listed species are documented window-kills, by definition they are very rare, and most buildings are not monitored. Consequently, the discovery of more endangered species is also likely to be rare.

All the complexities and confusion with applying the MBTA and ESA indicate federal legislation is currently of little use in addressing the window kill problem. Not in dispute is that laws requiring bird-safe windows will reduce the number of window kills. Another attempt to protect birds from windows at the federal level is a bill in the House of Representatives first introduced in 2010 by U.S. Representative Mike Quigley from Chicago. Originally titled the Federal Bird-Safe Building Act (H.B. 4797), its purpose was to force the use of bird-safe glass in remodeled and new government buildings. The hope was that such action, if passed, would stimulate private construction to follow and do the same. Neither this first nor any of its successors, H.R. 1643 in 2011, H.R. 2078 in 2013, H.R. 2280 in 2015, H.B. 2542 in 2017, and currently H.B. 919 in the 116th Congress (2018-2019) have had success in moving through the legislative process, let alone passing. The current bill has been incorporated in H.B. 2 and passed with several co-sponsors from both major U.S. political parties. Notwithstanding this initiative, there is currently no meaningful political will to address this life-and-death issue for birds at the federal level. To change this prospect, more effective education through more effective communication is required. As with DDT in the 1970s, garnering enough powerful voices will help ensure that our built environment is made safe for all free-flying wild birds.

At the federal level, the Canadians have been more courageous in using their equivalent to MBTA, the MBCA, and other environmental protection laws to address unintended window-kills. Using the EPA and SARA, an environmental non-profit legal firm, Ecojustice, led by attorney Albert Koehl, brought suit in 2012 against a property management company for birds killed striking the windows of a corporate center in Toronto (Podolsky versus Cadillac Fairview Corporation). Those on the side of the birds interpret the outcome of the case to be an environmental success. The court ruling established reflected light radiation to be responsible for creating an illusion that takes the lives of protected birds. Judge Melvyn Green, however, dismissed the case against Cadillac Fairview because they showed due diligence in retrofitting their offending windows with external film to mitigate continued bird casualties. The interpretation of environmental success was the expectation that other building managers at which window-kills are repeatedly documented will institute bird-safe practices to prevent their properties from being the target of future litigation. At least in this case, where great hope followed the ruling, the use of the legal system proved to be a far more powerful means of stimulating action to protect birds from windows than relying on voluntary efforts from so many diverse stakeholders.

Another federal litigation effort, with a far more successful outcome, took place in Cologne, Germany, in 2012. Using The Birds Directive agreement of the EU, the court ruled in a case that construction of a glass cube near and overlooking a bird conservation area was illegal because the proposed means to prevent bird-window collisions was not effective. The Der Bund fur Umwelt und Naturschutz Deutschland (BUND), the Association for Environment, and Nature Conservation Germany successfully sued the contractor. The decision required the installation of proven bird-safe glass with stripes instead of what the contractor judged to be a more aesthetically acceptable but ineffective collision deterrent.

Until federal governments use their power to protect birds throughout their sovereign domains, more promise and success in North America has occurred at the regional and local levels. State and provincial but mostly municipal government lawmakers have enacted a growing body of legislation to protect birds from windows. Most prominent among U.S. states is Minnesota, where state-funded building projects must comply with sustainability standards. Under those standards, all new construction and renovation requiring new windows must meet bird-safe guidelines. Although this regulation passed before the mirror-covered Viking Stadium was designed and built, the state exempted this construction, allowing previous regulations not requiring bird-safe building practices to apply. Similar state laws requiring bird-safe building practices are pending in Maryland, New York, Washington, District of Columbia, and Cook County, Illinois.

Specific mandatory ordinances and zoning regulations to prevent bird-window collisions at government and commercial buildings have been adopted by the following municipalities: Mountain View, Oakland, Richmond, San Francisco, and San Jose in California; Highland Park, Illinois, and Portland, Oregon; Markham and Toronto, Ontario. A similar *Bird Friendly Design Ordinance* (Chapter 13-150) is pending before Chicago's city council.

Voluntary recommendations have been formalized in Calgary, Alberta, and Vancouver, British Columbia; in the statewide California Green Building Code; in the California cities of Palo Alto and Sunnyvale; and the village of Barrington, Illinois. Each of these policies was drafted with the guidance of bird-safe building design guidelines published by the American Bird Conservancy, in cooperation by planning authorities and avian conservation advocates in the cities of Calgary, Markham, New York, Oakland, Portland, San Francisco, Toronto, and Vancouver.

Regarding recent legislative trends at all levels, as bird-safe sheet glass becomes more available and affordable, we can predict that human construction increasingly will be legally required to protect birds from windows the world over. One current progressive step is a grant to the Canadian Standards Association (CSA), an international organization, by the Ontario Ministry of Environment Conservation and Parks to develop *Bird Friendly Building Design* (CSA A460). This work recently has been completed and released to guide bird-safe building industry practices. These special bird-safe guidelines promise to be a model for bird-safe building construction worldwide.

Chapter 11

What Citizens Can do as Homeowners and Advocates

INDIVIDUAL CITIZENS WORKING ALONE, IN like-minded support groups, or as members of established conservation organizations can directly and indirectly act to protect birds from windows. If you have the will, time, and energy, do what you can to protect birds from this invisible killer. Of the 365 million to 988 million bird deaths that Loss and his colleagues estimated die annually by striking windows, in their 2014 article in the journal *Condor: Ornithological Applications*, 44% occurred at houses and multi-unit residences (1-3 stories), 56% at other residential and non-residential buildings (4-11 stories), and less than 1% at high rise structures of 12 or more stories. This study highlights how, collectively, individual homeowners can directly save hundreds of millions of bird lives by applying practical solutions to reduce or eliminate bird strikes at the windows of their dwellings. To those who have wondered about a muffled or startling thump or bang heard in their home, it was almost certainly a bird injured or killed outright hitting one of their windows, not a ghost attempting to communicate from the hereafter. Strings hanging over an offending picture window, such as Acopian BirdSavers, are cheap and easy to make and install.

Indirect action by individuals or with others may be employed to advocate for bird-safe buildings with government representatives and agencies at the federal, regional (provincial, state), and local levels, developers and managers of commercial and residential dwellings, and architects and film and glass manufacturers as building industry professionals. Organizations that monitor and evaluate environmentally responsible construction, such as the U.S. Green Building Council (USGBC), should be further encouraged to contribute to making sheet glass and plastic safe for birds. The USGBC currently uses an evaluation tool to assess environmentally responsible building, called Leadership in Energy Efficient Design (LEED). It recently added a Pilot 55 credit to address bird-safe construction, which is not yet a formal part of the assessment as an official credit. However, the content evaluating bird-safe products needs improvement to be more accurate and effective; that is, it is important to eliminate the risk that something may be assessed to be bird-safe when it is not. The highest LEED assessment for a newly constructed building is Platinum, but even a building that has attained this loftiest level of certification cannot be truly green if birds are killed flying into its windows.

Individual citizens can make a difference. Joan E. Joseph repeatedly found window-killed birds after they hit the glass walls of buildings in Independence National Historic Park, a National Park Service property in Philadelphia. Her compassionate concern, expressed in a heart-rending manner, regularly kept me informed of what she found there from 1995 to 1997. She brought what was to her an alarming number of bird deaths to the attention of the Park Service superintendent, which resulted in a written reply acknowledging the problem, describing several unsuccessful attempts to prevent it, and a claim that the lethal strikes have abated. Unfortunately, notwithstanding the superintendent's assurances that bird kills were now rare events, Joan continued to regularly find and report the dead, seeking more meaningful action. Starting in 2018, Amanda (Mandy) Meltz organized a lecture series and visited community and school leaders to explain the problem and request their help in protecting birds from windows in her Lower Merion Township, northwest of Philadelphia. Her collective outreach attracted local high school students to take up the cause of studying and supporting mitigation measures at their schools and elsewhere. Starting in 2014, encouraged by animal welfare scientist David Fraser at the University of British Columbia, executive director and CEO Ron Kagan and assistant curator of birds Bonnie Van Dam have committed to making all windows at the Detroit Zoological Society (DZS) in Royal Oak, Michigan, safe for free-flying wild birds, those that live near the zoo or pass through on migration. Zoos and aquaria have massive amounts of sheet glass in their buildings and outdoor exhibits, a fact probably not known to many. I learned of a remarkable event in 2006 when I was invited to speak about birds and windows at the Philadelphia Zoo, our nation's oldest zoo. Dr. Andy Baker shared a moving account with me about how

Kimya, a resident gorilla, witnessed and responded to a hummingbird that struck the clear glass wall of her enclosure. Kimya gently picked up what was almost certainly a Ruby-throated Hummingbird, our only resident hummingbird in the eastern U.S. With the lifeless body in her open palm, she slowly and carefully carried it to her keeper. The Detroit Zoo has made a commitment and taken action to make their glass safe, using signage to educate zoo visitors about the glass threat to birds, and offer for sale in the zoo gift shop proven tapes and films to prevent bird-window collisions. The DZS is a model for other zoos and aquaria on how to make their facilities safe for wild birds.

A recent initiative that holds great promise to educate and stimulate meaningful action was developed by Leigh Altadonna, Keith Russell, and Peter G. Saenger in southeastern Pennsylvania. They crafted a successful education grant proposal to the National Audubon Society (NAS), as a collaboration among Audubon Pennsylvania, Lehigh Valley Audubon Society, Wyncotte Audubon Society, and Muhlenberg College. The results of their extensive work is a 2019 Bird-Window Collision Presentation Toolkit from the Bird-Window Collision Working Group (BCWG). In conjunction with in-person instruction for those interested in becoming educators on the topic, the toolkit, available for use by Audubon societies throughout the U.S., other interested groups and individual activists such as Mandy Meltz, has great promise to reach those who believe they can make a difference in helping to make all windows safe for birds.

The content in the following sample letter is offered as a guide for individuals to introduce and request meaningful action on the bird-window issue wherever a suitable prospect can be identified. One example is the building managers of our federal, regional, and local park visitor centers. Typically, as was true for Independence National Historic Park in Philadelphia, these buildings are lavishly covered with glass that regularly kills the very birds the public come to see. Tragically, due to lack of knowledge or misunderstanding, bird feeders outside the windows of these centers are usually placed at distances beyond one meter (3.3 ft). Recall that attractants located beyond that distance from glass surfaces predictably attract more birds and result in increased deaths. Other prospective targets for sharing information about the fatal hazards of sheet glass for birds are elected officials (U.S.

representatives, senators, state governors, provincial premiers, mayors, council members, and building and zoning authorities), architects and landscape architects, developers, building managers, and other community leaders who can promote awareness and ideally facilitate action to retrofit existing lethal windows or make remodeled and new construction bird-safe upon occupation.

Chapter 11

[……DATE……]

[Individual or Organization
Address]

Reference: Bird-friendly Building Design

Dear [Individual or Organization],

Our human dwellings, from single-story houses to multi-story skyscrapers, are unintentionally killing valuable, innocent birds. Birds that have no voice or other means to protect themselves. These are species that have served the practical and spiritual needs of humans throughout recorded history. Scientific studies have repeatedly provided evidence showing that all birds behave as if windows are invisible to them. They fly into them, attempting to reach habitat and sky seen behind clear windows or reflected off the surface of sheet glass and plastic. The latest scientific estimates suggest that annual bird kills at windows range from 365 million to 988 million per year in the U.S., 16 million to 42 million in Canada, and billions worldwide. Even the lowest U.S. estimate is equivalent to the avian death toll from 1,215 Exxon Valdez oil-spill disasters. Yet most of the public knows little about this carnage. We should all know more about it, and you specifically, as someone in a position to do something about it. This is an environmental issue we humans can solve, and we must if we are to be responsible stewards of the one planet on which all life depends.

We know how to eliminate bird kills at windows, and there are many solutions that are either cost-neutral or investments that will come back to us via additional benefits. I am writing to you specifically in your capacity as [……………] to request that you apply your knowledge and authority to the limit that you can to make our human-built environment safe for bird life.

Two websites among many that can offer justification for devoting some of your valuable talent, time, and energy to this worthy cause are those prepared by the American Bird Conservancy (ABC; https://abcbirds.org/program/glass-collisions/) in the U.S., and the Fatal Light Awareness Program (FLAP; https://www.flap.org) in Canada.

Please contact me if you believe I can help you in any way related and appropriate to this important conservation issue for birds and people.

Sincerely yours,
[…….NAME……]
Contact Information]

Recommended minimum components of
MUNICIPAL ORDINANCE
or
ZONING CODE
for
BIRD-SAFE BUILIDNG DESIGN

Introduction, Rationale, Purpose, and Scope. Birds provide practical and spiritual services to humans. Scientific evidence, validated and repeatable, documents the fact that birds behave as if clear and reflective sheet glass and plastic are invisible to them. Birds sustain injury or death flying toward habitat and sky seen behind clear glass and reflected off the surface of windows. This regulation covers bird-friendly design at existing, remodeled, and new construction and is applicable to the building surface and site design. This regulation supersedes, and will define practice if there is any conflict with, previous regulations related to municipal building practices.

Definitions. [Include all specific appropriate structural and policy terminology, of which the following is an example.]

Glazing refers to sheet glass or plastic.

Markers are elements that make up a pattern uniformly covering the entire surface of the glass or plastic, consisting of elements that are visual (acid etch, ceramic frit, printed film, strings) or non-visual to humans (ultraviolet). Elements can be 2 mm (0.08 inch) or larger, with high contrast to unmarked areas.

The *2 x 4 Rule* defines markers covering a window where the elements are separated by 10 cm (4 inches) oriented in vertical columns, or 5 cm (2 inches) oriented in horizontal rows.

The *2 x 2 Rule* defines spacing between all pattern elements at 5 cm (2 inches).

Applicability. Existing, remodeled, and new structures are required to meet the described details of this regulation. Existing buildings can be treated with external films or other coverings on the outside surface (Surface #1) of panes. Clear panes can be treated with film or other coverings on any surface if markers are visible to birds and humans looking at the pane window from outside. Replacement or new windows installed in existing or new construction, respectively, will have visual or non-visual (UV) markers on the outside Surface #1 when covering a dark interior, or any surface if markers are visible to birds and humans looking at the window from outside.

Chapter 11

Architecture. Buildings should be designed to ensure birds are able to see and avoid sheet glass and plastic covering the outside surface.

All glazing preferably is designated bird-safe, or minimally consists of markers following the 2 x 4 Rule or 2 x 2 Rule to the height of 16 meters (52.5 ft).

Cover building external façades with shutters, louvers, grills, screening, netting if spacing between construction material follows the 2 x 4 Rule or 2 x 2 Rule spacing or less.

Exterior lighting to be shielded from above to light the building but not skyward.

Interior lighting control with sensors, and shielded from escaping through windows to exterior during darkness.

Among the prominent programs that invite individuals to participate in efforts to protect birds from windows are the Chicago Bird Collision Monitors in Chicago; the Fatal Light Awareness Program (FLAP) in Toronto; the New York City Audubon Society; the Bird-window Collision Working Group (BCWG) in Philadelphia and southeastern Pennsylvania; Portland Audubon in Portland, Oregon; and Lights Out D.C. in Washington, D.C.

Michael Mesure, co-founder and Executive Director of Fatal Light Awareness Program (FLAP), Toronto, Ontario, Canada.

The importance of effective leaders in any meaningful cause cannot be overstated. Some notable proponents of this cause are Michael Mesure of FLAP in Toronto, Rebekah Creshkoff of New York City Audubon, Randi Doeker of the Chicago Ornithological Society, Robbie Hunsinger and Annette Prince of Chicago Bird Collision Monitors, and Karen Cotton and Christine Sheppard of the American Bird Conservancy. They have made monumental efforts, but we need far more help informing the world about this terrible and unnecessary loss of bird life from window strikes and how to prevent them.

For those interested and moved to take action, details to guide you are provided in a 2015 *Biological Conservation* article by Scott Loss and his colleagues. They use bird-window collisions as a means of encouraging everyday citizens to help professional scientists make the case for promoting action to conserve bird life from the harmful actions of humans. These recommendations and those of other caring people remind us that we have the means to be good stewards of our Earth, but we need the will and the action to pull it off.

Chapter 12

Seeing the Invisible Threat – Overall Impact

Hospital emergency rooms treat thousands of injuries every year from people colliding with sheet glass in patio doors, exterior and interior glass walls, even shower doors. In all cases, the injured are distracted and not attentive to what at the time is an invisible barrier to them. Humans have died striking glass, but fatalities are rare. Most often there is a bruise and embarrassment, maybe a laceration, but no serious injury is sustained. People may not clearly see sheet glass, but they most often interpret and avoid it as a barrier because they are alerted to framing, a decal, or the context in which it is set such as clear or reflective panes making up a wall or doorway. By contrast, every individual of every species of bird is vulnerable to glass because of their inability to interpret windows as barriers.

Imprint of window strike on residential bedroom window. Photograph by Jamie Johns.

The health of any species with an interactive, structured society depends on its success at producing more of its kind than are lost. The leading members either directly contribute to prospective new leaders or support others who contribute leaders and non-leaders, keeping families and in turn their populations sustainable. Now imagine an invisible force indiscriminately culling, killing off, members of the population. Leaders and non-leaders alike are disappearing, but the population is hurt more severely by the loss of the strongest and most productive. Individual members do not possess the reasoned ability to identify and understand why their leaders and their kind are declining, let alone counter the damage. Individually nor collectively, they have no ability to be frightened or demoralized, just witnesses to reduced numbers and with it declining population health, based on fewer partners to increase and sustain stability. Enter a knowing alien species visitor who can identify and eliminate the invisible force. The visitors become the protectors because they can see what the affected species cannot. By applying protection before it is too late, the hapless and vulnerable are saved. Read windows as the invisible forces affecting birds; read knowing and caring humans as their saviors.

A review of the body of knowledge about this conservation issue includes: (a) quantitative studies revealing species, glass, building, and environmental conditions, (b) sustained injuries and cause of death, (c) level and composition of indiscriminate mortality as a conservation concern, (d) legislation to protect unintended victims, and (e) the means of mitigating and ideally eliminating the revolting number of casualties. Although we know birds are being killed globally by striking windows, is it possible that only a relatively small number of windows account for most of the mortality? If so, identifying and prioritizing the treatment of such select hazards, perhaps ground-level panes with associated attracting vegetation, can markedly reduce the carnage. Given the universal nature of how windows deceive birds, the prospect of identifying a select group of more dangerous windows is too hopeful, but still to be determined and worthy of further investigation.

It is recommended that future research include documenting which species are killed, in what numbers, and under what conditions at specific locations worldwide, using standard data collection methods and analyses. With improved measurements of population size and the mortality attributable to windows, it should be determined how the total accumulated number of individuals killed striking windows influences the population health of specific species and birds in general. Next, studies should determine the species-specific probability of surviving and recovering from the individual and collective injuries resulting from a window strike. Then research should decide on a standard method to measure the effectiveness of existing and yet-to-be developed solutions for transforming windows into barriers that birds will see and avoid. This would include defining what level of protection is bird safe; that is, reducing the risk of strikes by 50% to 90% or more. Leaders and architects need to be educated to effectively educate citizens and clients, respectively, about these unintended and unwanted deaths and, in so doing, encourage manufacturers and fabricators to offer additional bird-safe products for existing and new construction. More studies—separate, collaborative and collective—need to be encouraged in the life sciences, architecture, economics, law, psychology, sociology, and recreation to enhance the promise of saving more bird lives from windows.

Returning to hope is the best way forward. Invisible clear and reflective sheet glass and plastic is a lethal environmental hazard we can and should solve. The window threat is not as complex nor intractable as problems such as climate change. You have read how preventing bird-window collisions is justified for reasons that are ethical, moral, architectural, biological, economic, legal, psychological, and sociological. We humans created this unintentional and unwanted hazard for birds, and we must prepare and use bird-safe windows if we are to sustain the interdependence of species that is vital to the overall health of our planet.

There have been many successes in the movement to deal with the bird-window issue, and they are growing. No matter how modest our ability to protect birds from windows has been for the past four decades, progress over the past several years has been rapid, inspiring, and uplifting. There is still much to do. Ignoring the problem will not make it go away. We have the ability to fix it, but we still need the collective will. The birds still need and deserve a solution. We are their only hope, and I am confident we will make our windows safe for birds because it is the right thing to do.

Resources

American Bird Conservancy Website Collisions Section
https://abcbirds.org

American Bird Conservancy Guide to Bird Friendly Building Design
https://abcbirds.org/wp-content/uploads/2015/05/Bird-friendly-Building-Guide_2015.pdf

Best Practices for bird-window collision studies
Hager, S. B. and B. J. Cosentino. 2014. Surveying for bird-carcasses resulting from window collisions: A standardized protocol. PeerJ PrePrints 2014; 2. https://doi.org/10.7287/peerj.preprints.406v1.

Loss, S., S. S. Loss, T. Will, and P. P. Marra. 2014. Best practices for data collection in studies of bird-window collisions.
https://www.fws.gov/migratorybirds/pdf/management/Lossetal2014bestpracticesforwindowdata.pdf

Canadian Standards Association, Bird-friendly Building Design
store.csagroup.org

City of Toronto. 2007. Green development standard, bird-friendly development guidelines. City Planning, Toronto, Ontario, Canada.
http://www.toronto.ca/lightsout/pdf/development_guidelines.pdf

City of Toronto. 2016. Bird-friendly best practices glass.
http://www1.toronto.ca/City%20Of%20Toronto/City%20Planning/Environment/Files/pdf/B/BF%20Best%20Practices%20Glass_FinalAODA_Bookmarked.pdf

Fatal Light Awareness Program (FLAP)
http://flap.org/

Global Bird Collision Mapper (GBCM)
https://birdmapper.org/app/

Practical Deterrence Methods and Costs
http://pa.audubon.org/birds/windows

Resource Guide for Bird-friendly Building Design
http://audubonportland.org/files/hazards/bfbdd

Swiss Ornithological Society Bird-window Collisions and Light
https://vogelglas.vogelwarte.ch/assets/files/broschueren/Bird-friendly%20Building%20engl.pdf.

US Fish and Wildlife Service Best Practices Guide
https://www.fws.gov/migratorybirds/pdf/management/reducingbirdcollisionswithbuildings.pdf

Bird-Window Collision Prevention and Cost

A listing of retrofit bird-window collision deterrent methods and their relative costs courtesy of the Bird-Window Collision Working Group (BCWG) as of 2019.

Do-It-Yourself
Bird-window Collision Deterrent Methods and Relative Costs*

*Rough estimates; highly variable because of window sizes and configuration. We used a 36" high x 48" wide window for our calculations.

Acopian BirdSavers (www.birdsavers.com)
Easy to install, long-lasting, and highly effective. Simple and easy to make yourself. The BirdSavers website gives directions to build your own, or order them from Acopian BirdSavers. Approximately $2.98 materials cost,* or **25 cents per square foot**. if you build your own or $2.50 a square foot if you buy them premade.

Feather Friendly Bird Do-it-Yourself Tape
(www.conveniencegroup.com/featherfriendly/feather-friendly/)
Highly effective Do-It-Yourself Kit is now available from Feather Friendly® Technologies. Simple to install. Included instructions have you apply tape horizontally in a 2" x 2" pattern, but applying vertically using the 2" x 4" spacing seems effective and reduces material cost. Approximately $5 materials cost* or **40 cents per square foot**, if you use a 2" high x 4" wide pattern.

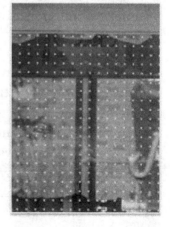

ABC BirdTape (http://www.abcbirdtape.org/)
Translucent tape, easily applied, easily removed. Lasts up to four years on outside surface. Space 4" apart if applied vertically or 2" apart if applied horizontally. Approximately $9.60 materials cost* or **80 cents per square foot** in the 4" vertical pattern.

CollidEscape (www.collidescape.org)
This perforated window film creates a solid appearance from the outside, but allows views from the inside and reduces glare and cooling costs while protecting birds. Available in plain white, stock colors, or you can have any image you want printed onto it. This is the type of material used for advertising on buses, etc.
Approximate $39–$79 materials cost* or **$3.25–$6.60 per square foot**.

Blinds
Blinds are found in many homes and can be used to discourage birds from flying into windows. When blinds are left partially open, a striped pattern is often visible from outside the window. The striped pattern can discourage collisions on windows that appear transparent or slightly reflective, by creating the appearance of a series of barriers.

Curtains and Shades
Curtains and shades can be used to discourage collisions on windows that appear transparent from the outside by simply closing them. Curtains and shades cannot reduce the formation of reflections on windows, however, and in some cases closing them may actually enhance the appearance of reflections.

Bird-Window Collision Prevention and Cost

Tempera Paint
Non-toxic tempera paint can be applied to the outer surface of a window to effectively eliminate both the appearance of transparency and the formation of reflections on a window. Because it is relatively inexpensive and easy to remove, it can be used to collision-proof windows in situations where more permanent or more expensive solutions may not be viable.

Screens (https://www.birdscreen.com/Index.htm) Fiberglass screening can also be mounted in front of windows to prevent bird collisions. If the screening is mounted at least 5 inches or more in front of the glass with enough tension, it will allow most birds that fly into it to bounce off without suffering the fatal or life-threatening injuries that could result from striking the hard glass surface directly. Screening is less effective when used on larger windows that require it to be stretched past the point where it can be maintained with enough

tension to remain in a fairly rigid position. **Approximately $24–$36 material cost* or $1.83–$2.69 cents per square foot.**

Designed and produced by the
Bird-Window Collision Group *(BWWG),*
a collaboration between The Acopian Center for Ornithology, Muhlenberg College, Lehigh Valley Audubon Society, Wyncote Audubon Society, and Audubon Pennsylvania.

Reproduction and distribution allowed with credit given to the **Bird-Window Collision Group.**

Bird-Window Collision Species by Geography

List of number of bird species known to strike windows by geographic region and country. The list identifies species records, and is presented to encourage additional reporting from locations with little or no documentation.

North America
Canada-198
Mexico-87
United States- 278

Central America
Belize-3
Costa Rica-268
El Salvador-4
Guatemala-2
Panama-1

South America
Argentina-15
Brazil-150
Chile-2
Colombia-49
Ecuador-3
Peru-7

Caribbean and Vicinity
Aruba-5
Bahamas-2
Bermuda-23
British Virgin Islands-14
Cuba-10
Curacao-1
Dominica-6
Dominican Republic-3
Guadeloupe-10
Haiti-1
Jamaica-2
Puerto Rico-17

Europe
Armenia-5
Austria-48
Belarus-1
Belgium-13
Czech Republic-5
Denmark-9
Finland-36
France-44
Germany-123
Gibraltar-1
Iceland-16
Italy-54
Luxembourg-36
Montenegro-1

Netherlands-91
Norway-51
Poland-54
Portugal-11
Romania-2
Russia-11

Slovenia-15
Spain-45
Sweden-89
Switzerland-137
Ukraine-4
United Kingdom-125

Middle East
Israel-8

Turkey-3

Africa
Kenya-5
Madagascar-2

South Africa-68
Zambia- 1

Asia
Cambodia-1
China-47
India-12
Indonesia-2
Japan-79
Malaysia-1

Nepal-4
Singapore-36
Republic of Korea-136
Taiwan-1
Timor-Leste-1

Australian Region
Australia-207
New Zealand-4

Papa New Guinea-7
Tasmania-5

1,348 documented species as bird-window collisions in 73 counties and territories.

*This list is out-of-date almost as soon as it is printed. We are continuously making new contacts around the world, adding to our database of bird-window collisions. Any species not already on our list will eventually be on it if it occurs near unprotected sheet glass or plastic.

Visit http://aco.muhlenberg.edu for current list.

Bird Species Known to Strike Windows

World list of bird-window Collisions: 1348 speces.

*This list is out of date almost as soon as it is printed. We are continuously making new contacts around the world, adding to our database of bird-window collisions. Any species not already on our list will eventually be on it if it occurs near unprotected sheet glass or plastic.

Visit http://aco.muhlenberg.edu for current list.

Bibliography

Able, K. P., and M. A. Able. 1995. Manipulations of polarized skylight calibrate magnetic orientation in a migratory bird. Journal of Comparative Physiology and Sensory Neural and Behavioral Physiology 177:351-356. doi:10.1007/BF00192423.

Acopian BirdSavers. 2014. Available online: http://www.birdsavers.com.

Acopian Center for Ornithology. 2019. World list of bird species documented striking windows. Online at http://aco.muhlenberg.edu.

Adalsteinsson, S. A., J. J. Buler, J. L. Bowman, V. D'Amico, Z. S. Ladin, and W. G. Shriver. 2018. Post-independence mortality of juveniles is driven by anthropogenic hazards for two passerines in an urban landscape. Journal of Avian Biology 49. https://doi.org/10.111/jav.01555.

Adams, C. A., A. Blumenthal, E. Fernandez-Juricic, E. Bayne, and C. C. St. Clair. 2019. Effect of anthropogenic light on bird movement, habitat selection, and distribution: a systematic map protocol. Environmental Evidence 8(Supppl 1):13. https://doi.org/10.1186/s13750-019-0155-5.

Adrian, M. 2014. The mincing mockingbird, Guide to Troubled birds. Blue Rider Press, A Member of Penguin Group (USA), New York, New York.

Agudel-Alvarez, L. 2006. Colision de aves contra los ventanales del campus de la Universidad Javeriana, Sede Bogota. Alternativas de mitigacion. Bogota: Universidad Javeriana.

Agudel-Alvarez, L., J. Moreno-Velasquez, N. Ocampo-Penuela. 2010. Colisiones de aves contra ventanales en un campus universitario de Bogota, Colombia. Ornitologia Colombiana 10:3-10.

Ahiemstra, M., E. Kdlabola, and E. L. O'Brien. 2020. Factors influencing bird-window collisions in Victoria British Columbia. Northwestern Naturalist 101:27-33.

Algeo, M. 2009, Harry Truman's excellent adventure: The true story of a great American road trip. Chicago Review Press, Chicago, Illinois, USA.

Allen, J. A. 1880. Destruction of birds by light-houses. Bulletin of the Nuttall Ornithological Club 5:131-138.

Alonso, J. C., J. A. Alonso, and R. Munoz-Pulido. 1994. Mitigation of bird collisions with transmission lines through groundwire marking. Biological Conservation 67:129-134.

American Birding Association. 2007. ABA Checklist. Available from http://. americanbirding.org/checklist/index.html.

American Ornithologists' Union. 1975. Report of the ad hoc committee on scientific educational use of wild birds. Auk 92 (Suppl.):1A-27A.

American Ornithologists' Union. 1983. Check-list of North American birds, 6th Edition. American Ornithologists' Union, Baltimore, Maryland, USA.

Arbab, M. and J. J. Finley. 2010. Glass in architecture. International Journal of Applied Glass Science 1:118-129. doi:10.1111/j2041-1294.2010.00004.x.

Arnold, T. W. 2010. Uninformative parameters and model selection using Akaike's information criterion. Journal of Wildlife Management 74:1175-1178.

Arnold, T. W. and R. M. Zink. 2011. Collision mortality has no discernible effect on population trends of North American birds. PLoS One 6, e24708.

Audubon Minnesota. 2010. Bird-safe building guidelines. Saint Paul, Minnesota, USA.

Ausprey, I. J., A. D. Rodewald. 2011. Postfledging survivorship and habitat selection across a rural-to-urban landscape gradient. Auk 128:293-302.

Avery, M. L., P. F. Springer, and J. L. Cassel. 1976. The effects of a tall tower on nocturnal bird migration: A portable ceilometer study. Auk 93:1-12.

Avery, M. L. 1979. Review of avian mortality due to collisions with man-made structures. U.S. Fish and Wildlife Service, 11pp. http://digitalcommons.unl.edu/egi/viewcontent.egi?article=1001&context=icwdmbirdcontrol.

Avolio, M. L., D. E. Pataki, T. L. E. Trammell, and J. Endter-Wada. 2018. Biodiverse cities: The nursery industry, homeowners, and neighborhood differences drive urban tree composition. Ecological Monographs 88:259-276. DOI:10.1002/ecm.1290.

Aymi, R., Y. Gonzalez, T. Lopez, and O. Gordo. 2017. Bird-window collisions in a city on the Iberian Mediterranean coast during autumn migration. Revista Catalana d' Ornithologia 33:17-28.

Baily, W. L. 1990. Migration data on city hall tower. Abstract of the Proceedings of the Delaware Valley Ornithological Club 1898-1899: 15-19.

Baines, D., and R. W. Summers. 1997. Assessment of bird collisions with deer fences in Scottish forests. Journal of Applied Ecology 34:941-948.

Baird, S. F., T. M. Brewer, and R. Ridgway. 1874a. A history of North American birds: land birds, Volume I. Little, Brown and Company, Boston, Massachusetts, USA.

Baird, S. F., T. M. Brewer, and R. Ridgway. 1874b. A history of North American birds: land birds, Volume III. Little, Brown and Company, Boston, USA.

Balogh, A. P., T. B. Ryder, and P. P. Mara. 2011. Population demography of Gray Catbirds in the suburban matrix: sources, sinks and domestic cats. Journal of Ornithology DOI: 10.1007/s10336-011-0648-7.

Banks, R. C. 1976. Reflective plate—hazard to migrating birds. Bioscience 26: 414.

Banks, R. C. 1979. Human related mortality of birds in the United States. U.S. Fish and Wildlife Service Special Report 215: 1-16.

Banks, R. C., M. H. Clench, and J. C. Barlow. 1973. Bird collections in the United States and Canada. Auk 90:136-170.

Barber, D. 2014. The third plate: Field notes on the future of food. The Penguin Press, New York, New York.

Barton, C. M., C. S. Riding, and S. R. Loss. 2017. Magnitude and correlates of bird collisions at glass bus shelters in an urban landscape. PLoS ONE 12(6):e0178667. https://doi.org/10.1371/journal.pone.0178667.

Basilio, L. G., D. J. Moreno, and A. J. Piratelli. 2020. Main causes of bird-window collisions: a review. Annals of the Brazilian Academy of Sciences 92(1):11 pp.

Bator, R. J., and R. B. Cialdini. 2000. New ways to promote proenvironmental behavior: The application of persuasion theory to the development of effective proenvironmental public service announcements. Journal of Social Issues 56:527-541. DOI:10.1111/0022-4537.00182.

Batty, M. 2008. The size, scale, and shape of cities. Science 319:769-771.

Bauer, E. W. 1960. Vogeltod an Glaswanden. Aus der Heimat 68:58-60.

Bauer, S. and B. J. Hoye. 2014. Migratory animals couple biodiversity and ecosystem functioning worldwide. Science 344. doi:10.1126/science.1242552.

Bayne, E. M., C. A. Scobic, and M. Rawson-Clark. 2012. Factors influencing the annual risk of bird-window collisions at residential structures in Alberta, Canada. Wildlife Research 39: 583-592.

Beason, R. 2012. Avian visual perception A literature review. USDA/APHIS/WS National Wildlife Research Center, Sandusky, Ohio, USA. Contracted report for Electric Power Research Institute (EPRI), Palo Alto, California, USA.

Beebe, W. 1949. High jungle. Duell, Sloan and Pearce, New York, USA.

Belcher, R. N., K. R. Sadanandan, E. R. Goh, J. Y. Chan, S. Menz, and T. Schroepfer. 2019. Vegetation on and around large-scale buildings positively influences native tropical bird abundance and bird species richness. Urban Ecosystems 22:213-225.

Bennett, A. T. D. and I. C. Cuthill. 1994. Ultraviolet vision in birds: what is its function? Vision Research 34: 1471-1478.

Bennett, A. T., I. C. Cuthill, J. C. Partridge, and E. J. Maier. 1996. Ultraviolet vision and mate choice in zebra finches. Nature 380: 433-435.

Bent, A. C. 1940. Life histories of North American cuckoos, goatsuckers, hummingbirds and their allies. United States National Museum Bulletin 176, Washington D.C.

Bent, A. C. 1949. Life histories of North American thrushes, kinglets, and their allies. United States National Museum Bulletin 196, Washington D.C.

Bent, A. C. 1968. Life histories of North American cardinals, grosbeaks, buntings, towhees, finches, sparrows, and their allies. United States National Museum Bulletin 237, Washington D.C.

Berenstein, N. 2015. Deathtraps in the flyways: electricity, glass and bird collisions in urban North America, 1887-2014, in Cosmopolitan Animals, K. Nagai, K. Jones, D. Landry, M. Mattfeld, C. Rooney, and C. Sleigh, Editors, Palgrave Macmillan, U.K., 79-92.

Bernard, R. F. 1966. Fall migration – western Great Lakes region. Audubon Field Notes 20:45-46, 50-53.

Bessinger, S. R., and D. R. Osborne. 1982. Effects of urbanization on avian community organization. Condor 84:75-83.

Best, J. 2001. Damned lies and statistics: untangling numbers from the media, politicians, and activists. University of California Press, Berkley, California, USA.

Best, J. 2008. Birds – dead and deadly: Why numeracy needs to address social construction. Numeracy 1:1-14. DOI: 10.5038/1936-4660.1.1.6.

Bevanger, K. 1998. Biological and conservation aspects of bird mortality caused by electricity power lines: A review. Biological Conservation 86:67-76.

Bhagavatula, P., C. Claudianos, M. R. Ibbotson, and M. V. Srinivasan. 2009. Edge detection in landing Budgerigars (*Melopsittacus undulates*). PLoS ONE 4(10):e7301. doi:10.1371/journal.pone.0007301.

Bhagavatula, P., C. Claudianos, M. R. Ibbotson, and M. V. Srinivasan. 2011. Optical flow cues guide flight in birds. Current Biology 21:1794-1799. doi:10.1016/j.cub.2011.09.009.

Biagi, N. 2019. Breeding season bird mortality from window collisions: Comparing species-specific abundance with mortality rates. MURAJ 1:1-10. Z.umn.edu/MURAJ.

Bird, D. M. 2000. A reflection on window kills: Glass windows are having a major impact on birds. Bird Watcher's Digest 22:76-84.

Bird, D. M. 2012. The impact of impacts. Bird Watcher's Digest 34:108-113.

Bird Screen. 2014. Available online: http://www.birdscreen.com (accessed on 11 March 2014).

BirdLife International. 2000. Threatened birds of the world. Lynx Edicions and BirdLife International, Barcelona, Spain, and Cambridge, UK. Updates available from http://www.bird life.org/action/science/species/global_species_programme/whats_new.html (accessed 11 March 2014).

Bishop, C. A., and J. M. Brogan. 2013. Estimates of avian mortality attributed to vehicle collisions in Canada. Avian Conservation and Ecology 8:2.

Black, C. A. 1922. Some bird notes from central and western Nebraska. Wilson Bulletin 34:43.

Blackwell, B. F., E. Fernandez-Juricic, T. W. Seamans, and T. Dolan. 2009. Avian visual system configuration and behavioural response to object approach. Animal Behaviour 77:673-684. doi:10.1016/j.anbehav.2008.11.017.

Blain, A. W. 1948. On the accidental death of wild birds. Jack-Pine Warbler 26:59-60.

Blair, R. B. 1996. Land use and avian species diversity along an urban gradient. Ecological Applications 6:506-519.

Blair, R. 2004. The effects of urban sprawl on birds at multiple levels of biological organization. Ecology and Society 9:2.

Blair, R. B., and E. M. Johnson. 2008. Suburban habitats and their role for birds in the urban-rural habitat network: points of local invasion and extinction? Landscape Ecology 23:1157-1169.

Blancher, P. J. 2013. Estimated number of birds killed by house cats (*Felis catus*) in Canada. Avian Conservation Ecology, 8, doi:10.5751/ACE-00557-080203.

Blem, C. and B. Willis. 1998. Seasonal variation of human-caused mortality of birds in the Richmond area. The Raven 69:3-8.

Blest, A. D. 1957. The function of eyespots in the Lepidoptera. Behaviour 11:209-256.

Blode, J., and F. Girandola. 2016. Revealing the elusive effects of vividness: A meta-analysis of empirical evidences assessing the effect of vividness on persuasion. Social Influence 11:111-129. DOI:10.1080/15534510.2016.1157096.

Blokpoel, H. 1976. Bird hazards to aircraft. Clarke, Irwin and Company, Toronto, Ontario, Canada.

Bocetti, C. I. 2011. Cruise ships as a source of avian mortality during fall migration. Wilson Journal of Ornithology 123:176-178.

Bolshakov, C. V., M. V. Vorotkov, A. Y. Sinelschikova, V. N. Bulyuk, and M. Griffiths. 2010. Application of the Optical-Electronic Device for the study of specific aspects of nocturnal passerine migration. Avian Ecological Behavior 18:23-51.

Bolshakov, C. V., V. N. Bulyuk, A. Y. Sinelschikova, and M. V. Vorotkov.2013. Influence of the vertical light beam on numbers and flight trajectories of night-migrating songbirds. Avian Ecological Behavior 24:35-49.

Bonter, D. N., S. A. Gauthreaux, Jr., and T. M. Donovan. 2009. Characteristics of important stopover locations for migrating birds: remote sensing with radar in the Great Lakes basin. Conservation Biology 23:440-448.

Borden, W. C., O. M. Lockhart, A. W. Jones, and M. S. Lyons. 2010. Seasonal, taxonomic, and local habitat components of bird-window collisions on an urban university campus in Cleveland, OH. Ohio Journal of Science 110:44-52.

Bostrom, J. E., M. Dimitrova, C. Canton, O. Hastad, A. Qvarnstrom, and A. Odeen. 2016. Ultra-rapid vision in birds. PLoS One 11 (3), e0151099. https://doi.org/10.1371/jounal.pone.0151099.

Boyle, S. A., and F. B. Sampson. 1985. Effects of nonconsumptive recreation on wildlife: a review. Wildlife Society Bulletin 13:110-116.

Bracey, A. M. 2011. Window related avian mortality at a migration corridor. Master's Thesis, University of Minnesota, Duluth, Minnesota, USA.

Bracey, A. M., M. A. Etterson, G. J. Niemi, and R. F. Green. 2016. Variation in bird-window collision mortality and scavenging rates within an urban landscape. The Wilson Journal of Ornithology 128:355-367.

Bradley, M. 1975. A study and analysis of man-made navigational hazards to birds in the vicinity of Richmond, Indiana. [Unpublished Manuscript, Earlham College, Richmond, Indiana].

Bradley, C. A., and S. Altizer. 2007. Urbanization and the ecology of wildlife diseases. Trends in Ecology and Evolution 22:95-102.

Breithaupt, M. A. K. Davis, and R. Hall. 2013. A preliminary survey of birds killed by window collisions in Georgia based on museum specimens Oriole 77:9-17.

Brisque, T., L. A. Campos-Silva, and A. J. Piratelli. 2017. Relationship between bird-of-prey decals and bird-window collisions on a Brazilian university campus. Zoologia 34:1-8.

Brittingham, M. C., and S. A. Temple. 1986. A survey of avian mortality at winter feeders. Wildlife Society Bulletin 14:445-450.

Brittingham, M. C., and S. A. Temple. 1992. Does winter bird feeding promote dependency? Journal of Field Ornithology 63:190-194.

Brown, B. B., E. Kusakabe, A. Antonopoulos, S. Siddoway, and L. Thompson. 2019. Winter bird-window collisions: Mitigation success, risk factors, and implementation challenges. PeerJ 7:e7620. http://doi.org/10.7717/peerj.7620.

Brown, B. B., L. Hunter, and S. Santos. 2020. Bird-window collisions: different fall and winter risk and protective factors. PeerJ 8:e9401. http://doi.org/10.7717/peerj.9401.

Brown, H. and S. Caputo. 2007. Bird-safe building guidelines. New York City Audubon Society, New York, New York, USA.

Bruijns, M. F. Morzer and L. J. Stwerka. 1961. Het doodvliegen van vogels tegen ramen. De Levende Natuur 64:253-257.

Buck, C. F. 1936. Intemperate robins. Nature Notes 1:129-130.

Buer, F., and M. Regner. With the 'spider's web effect' and UV-absorbing material against bird-death on transparent and reflecting panes. 2002. Zeitschrift fur Vogelkunde und Naturschutz in Hessen – Vogel und Umwelt 13:31-41.

Bulyuk, V. N., C. V. Bolshakov, A. Y. Sinelschikova, and M. V. Vorotkov.2014. Does the reaction of nocturnally migrating songbirds to the local light source depend on backlighting of the sky? Avian Ecological Behavior 25:21-26.

Bureau of the Census. 1976. Statistical abstract of the United States: 1976. 97th Edition, Washington, D.C., USA.

Burkhardt, D. 1982. Birds, berries and UV. Naturwissenschaften 69: 153-157.

Burnham, K. P., and D. R. Anderson. 2002. Model selection and multimodel inference, 2nd Ed. Springer-Verlag, New York, NY USA.

Burns, K. J. and A. J. Shultz. 2012. Widespread cryptic dichromatism and ultraviolet reflectance in the largest radiation of Neotropical songbirds: implications of accounting for avian vision in the study of plumage evolution. Auk 129:211-221.

Burr, Brooks M. and David M. Current. 1974. The 1972-1973 Goshawk invasion of Illinois. Transactions of the Illinois Academy of Science 67:175-179.

Burtt, E. H., Jr. and W. E. Davis, Jr. 2013. Alexander Wilson: The Scot who founded American ornithology. The Belknap Press of Harvard University Press, Cambridge, Massachusetts, USA.

Butcher, G. S., D. K. Niven, A. O. Panjabi, D. N. Pashley, and K. V. Rosenberg. 2007. The 2007 WatchList for United States Birds. American Birds 61: 18-25.

Bibliography

Butler, M. 1986. Observations of bird mortality at office towers in downtown Toronto: 1976-1985. Course 396c, [Unpublished Manuscript], Trent University, Peterborough, Ontario, Canada.

Cabrera-Cruz, S. A., J. A. Smolinsky, and J. J. Buler. 2018. Light pollution is greatest within migration passage areas for nocturnally-migrating birds around the world. Scientific Reports 8:3261. doi:10.1038/s41598-018-21577-6.

Cabrera-Cruz, S. A., J. A. Smolinsky, K. P. McCarthy, and J. J. Buler. 2019. Urban areas affect flight altitudes of nocturnally migrating birds. Journal of Animal Ecology. doi:10.1111/1365-2656.13075.

Calvert, a. M., C. A. Bishop, R. D. Elliot, e. A. Krebs, T. M. Kydd, C. S. Machtans, and G. J. Robertson. 2013. A synthesis of human-related avian mortality in Canada. Avian Conservation and Ecology 8:11.

Campedelli, T., G. Londi, S. Cutini, C. Donati, and G. T. Florenzano. 2014. Impact of noise barriers on birds. A case study along a Tuscany highway. Avocetta 38:37-39.

Canadian Standards Association. 2019. CSA A460:19 Bird-friendly building design. CSA Group. https://store.csagroup.org/ccrz__ProductDetails?viewState=DetailView&cartID=&portalUser=&store=&cclcl=en_US&sku=CSA%20A460%3A19.

Carpenter, F. and H. B. Lovell. 1963. Bird casualties near Magnolia, Larue County. Kentucky Warbler 39:19-21.

Carpentier, G. 2013. Windows and birds. Newsletter of the Ontario Field Ornithologists 31:1-4.

Carson, R. 1962. Silent spring. Houghton Mifflin Company, Boston, Massachusetts. USA.

Carver, E. 2013. Birding in the United States: a demographic and economic analysis Addendum to the 2011 National Survey of Fishing, Hunting, and Wildlife-Associated Recreation, Report 2011-1, U. S. Department of the interior, U.S. Fish and Wildlife Service, Division of Economics, Arlington, Virginia, USA.

Chace, J. F., and J. J. Walsh. 2006. Urban effects on native avifauna: a review. Landscape Urban Plan 74:46-69.

Chamberlain, B. R. 1959. Roadblock, an experiment. Chat 23:32-33.

Chicago Ornithological Society. 2014. Available online: http://chicagobirder.org/conservation/birds-building-collisions.

Chin, S. 2016. Investigating the effects of urban features on bird window collisions. Master of Science Thesis, York University, Toronto, Ontario.

Chubb, Kit. 2004. A study of 397 window collisions. Avian Care and Research Foundation, Verona, Ontario, Canada [Unpublished Manuscript].

Churcher, P. B. and J. H. Lawton. 1987. Predation by domestic cats in an English-village. Journal of Zoology 212:439-455.

Ciach, M., and A. Frohlich. 2017. Habitat type, food resources, noise and light pollution explain the species composition, abundance and stability of a winter bird assemblage in an urban environment. Urban Ecosystems 20:547-559. DOI:10.1007/s11252-016-0613-6.

City of Calgary. 2011. Bird-friendly urban design guidelines. Land Use Planning and Policy. Calgary, Alberta, Canada.

City of Chicago. 2007. Bird-safe building: design guide for new construction and renovation. Chicago, Illinois, USA.

City of San Francisco. 2011. Standards for bird-safe buildings. Planning Department, San Francisco, California, USA.

City of Toronto. 2007. Bird-friendly development Guidelines. Green Development Standard, City Planning, Toronto, Ontario, Canada.

City of Toronto. 2016. Bird-friendly best practices glass. City Planning, Toronto, Ontario, Canada.

City of Vancouver. 2015. Bird strategy. Vancouver, British Columbia, Canada.

Clergeau, P., J. P. L. Savard, G. Mennechez, and G. Falardeau. 1998. Bird abundance and diversity along an urban–rural gradient: a comparative study between two cities on different continents. Condor 100:413-425.

Codoner, N. A. 1995. Mortality of Connecticut birds on roads and at buildings. Connecticut Warbler 15:89-98.

CollidEscape. 2014. Available online: http://www.collidescape.org (accessed on 11 March 2014).

Collins, K. A., D. J. Horn. 2008. Bird-window collisions and factors influencing their frequency at Millikin University in Decatur, Illinois. Transactions of the Illinois State Academy of Science 101 (Supplement): 50.

Convenience Group Incorporated. 2015. Feather friendly bird deterrent technology. Online at http://www.conveniencegroup.com/featherfriendly/feather-friendly.

Corcoran, L. M. 1999. Migratory Bird Treaty Act: strict criminal liability for non-hunting caused bird deaths. Denver University Law Review 77: 315-358.

Costanza, R., R. d'Arge, R. de Groot, S. Farber, M. Grasso, B. Hannon, K. Limburg, S. Naeem, R. V. O'Neill, J. Paruelo, R. G. Raskin, P. Sutton, and M. van den Belt. 1997. The value of the world's ecosystem services and natural capital. Nature 387:253-260. doi:10.1038/387253a0. https://www.nature.com/articles/387253a0.

Crawford, R. L. 1981a. Bird kills at a lighted man-made structure: often on nights close to a full moon. American Birds 35:913-914.

Crawford, R. L. 1981b. Weather, migration and autumn bird kills at a north Florida TV tower. Wilson Bulletin 93:189-195.

Crawford, R. L., and R. T. Engstrom. 2001. Characteristics of avian mortality at a north Florida television tower: a 29-year study. Journal of Field Ornithology 72:380-388.

Cruickshank, A. D. and H. G. Cruickshank. 1958. 1001 questions answered about birds. General Publishing Company, Toronto, Ontario, Canada.

Culver, D. 1915. Mortality among birds at Philadelphia, May 21-22 1915. Cassinia 19:33-37.

Cupul-Magana, F. G. 2003. Nota sobre colisiones de aves en las ventanas de edificios universitarios en Puerto Vallarta, Mexico. Huitzil 4:17-21.

Cusa, M., D. A. Jackson, M. Mesure. 2015. Window collisions by migratory bird species: urban geographical patterns and habitat associations. Urban Ecosystems 18:1-20 DOI 10.1007/s11252-015-0459-3.

Daily, G. C., Ed. 1997. Nature's services: Societal dependence on natural ecosystems. Island Press, Washington, D.C.

Dakin, R., T. K. Fellows, D. L. Altshuler. 2016. Visual guidance of forward flight in hummingbirds reveals control based on image features instead of pattern velocity. Proceedings of the National Academy of Sciences 113:8849-8854.

Dawson, g. A. and P. L. Dalby. 1973. A goshawk-thermopane encounter. Jack-Pine Warbler 51:128.

DeCandido, R. 2005a. Dancing in the moonlight: nocturnal bird migration from the top of the Empire State Building. Winging It 19:1-5.

DeCandido, R. 2005b. Night moves: nocturnal bird migration from the top of the Empire State Building. Birder's World 19:6-7.

DeCandido, R. and D. Allen. 2006a. Spring 2004 visible night migration of birds at the Empire State Building, New York City. The Kingbird 56(3):199-210.

DeCandido, R. and D. Allen. 2006b. Nocturnal hunting by peregrine falcons at the Empire State Building, New York City. Wilson Journal of Ornithology 118:53-58.

Deck, R. S. 1941. Pageant in the sky. Dodd, Mead and Company, New York, USA, in Peterson, R. T. 1957. The bird watcher's anthology. Harcourt, Brace, and Company, New York, USA, 90-93.

DeVault, T. L., O. E. Rhodes, Jr., and J. A. Shivik. 2003. Scavenging by vertebrates: behavioral, ecological and evolutionary perspectives on an important energy transfer pathway in terrestrial ecosystems. Oikos 102:225-234.

Diederich, J. 1977. Vogelverluste an Glasflaechen des Athenaeums in Luxemburg. Regulus 12:137-139.

Diehl, R. H. 2013. The airspace is habitat. Trends in Ecology and Evolution 28:1-3.

Dolbeer, R. A. 2006. Height distribution of birds recorded by collisions with civil aircraft. Journal of Wildlife Management 70:1345-1350.

Donald, P., R. Green, and M. Heath. 2001. Agricultural intensification and the collapse of Europe's farmland bird populations. The Royal Sociey 268:25-29.

Drewitt, A. L., R. H. W. Langston. 2008. Collision effects of wind-power generators and other obstacles on birds. Annual New York Academy of Science 1134:233-266.

Dunbar, R. J. 1949. Birds colliding with windows. Migrant 20:12-15.

Dunn, E. H. 1993. Bird mortality from striking residential windows in winter. Journal of Field Ornithology 64(3): 302-309.

Ebersole, R. and M. Brandon. 2015. Bird vs. building: Capturing the effects of steel and glass on birds that can't see them. Audubonj 117:32-37.

Elmhurst, K. S., and K. Grady. 2017. Fauna protection in a sustainable university campus: Bird-window collision mitigation strategies at Temple University. In: Handbook of theory and practice of sustainable development in higher education. Cham: Springer, 69-82.

Elmore, J. A., S. B. Hager, B. J. Cosentino, T. J. O'Connell, C. S. Riding, M. L. Anderson, M. H. Bakermans, T. J. Boves, D. Brandes, E. M. Butler, M. W. Butler, N. L. Cagle, R. Calderon-Parra, A. P. Capparella, A. Chen, K. Cipollini, R. L. Curry, J. J. Dosch, K. L. Dyson, E. E. Fraser, R. A. Furbush, N. D. G. Hagemeyer, K. N. Kopfensperger, D. Klem, Jr., E. A. Lago, A. S. Lahey, C. S. Machtans, J. M. Madosky, T. J. Maness, K. J. McKay, S. B. Menke, N. Ocampo-Penuela, R. Ortega-Alvarez, A. L. Pitt, A. Puga-Caballero, J. E. Quinn, A. M. Roth, R. T. Schmitz, J. L. Schnurr, M. E. Simmons, A. D. Smith, C. W. Varian-Ramos, E. L. Walters, L. A. Walters, J. T. Weir, K. Winnett-Murray, I. Zuria, J. Vigliotti, and S. R. Loss. 2020. Correlates of bird collisions with buildings across three North American countries. Conservation Biology . DOI:10.1111/cobi.13569.

Emlen, J. T., Jr. 1963. Determinants of cliff edge and escape responses in Herring Gull chicks in nature. Behaviour 22:1-15.

Endler, J. A., and P. W. Mielke, Jr. 2005. Comparing entire colour patterns as birds see them. Biological Journal of the Linnean Society 86:405-431.

Erickson, W. P., G. D. Johnson, M. D. Strickland, D. P. Young, Jr., K. J. Sernka, and R. E. Good. 2001. Avian collisions with wind turbines: a summary of existing studies and comparisons to other sources of avian collision mortality in the United States. National Wind Coordinating Committee, Washington, D.C.

Erickson, W. P. G. D. Johnson and D. P. Young. 2005. A summary and comparison of bird mortality from anthropogenic causes with an emphasis on collisions. USDA Forest Service General Technical Report PSW-GTR-191.

Errington, P. L. 1946. Predation and vertebrate populations. Quarterly Review Biology 21:144-147.

Erritzoe, J., T. D. Mazgajski, and L. Rejit. 2003. Bird casualties on European roads – A review. Acta Ornithologica 38:77-93.

Evans, A. M. 1976. Reflective glass. Bioscience 26:596.

Evans, W. R., Y. Akaski, N. S. Altman, and A. M. Manville, II. 2007. Response of night-migrating songbirds in cloud to colored and flashing light. North American Birds 60:476-488.

Evans Ogden, L. J. 1996. Collision Course: the hazards of lighted structures and windows to migrating birds. World Wildlife Fund Canada and Fatal Light Awareness Program, Toronto.

Evans Ogden, L. J. 2002. Summary report on the Bird Friendly Building program: effect of light reduction on collision of migratory birds. Special Report for the Fatal Light Awareness Program (FLAP).

Evans Ogden, L. J. 2014. Does green building come up short in considering biodiversity? Bioscience 64:83-89.

Faeth, S. H., P. S. Warren, E. Shochat, and W. A. Marussich. 2005. Tropic dynamics in urban communities. BioScience 55:399-407.

Fatal Light Awareness Program. 2014. Available online: http://www.flap.org (accessed 11 March 2014).

Faver, A. R. 1968. Tinklebells warn birds. Chat 32:9.

Fedun, I. 2000. The built environment: a bird's-eye view. Canadian Architect 45:48-49.

Fiedler, W. and H.-W. Ley. 2013. Ergebnisse von Flugtunnel-Tests im Rahmen der Entwicklung von Glasscheiben mit UV-Signatur zur Vermeidung von Vogelschlag. Vogel-schutz 49/50: 115-134.

Findlay, S. 2011. This one's for the birds. Maclean's 124:68-70.

Fink, L. C. 1970. Birds in downtown Atlanta – 1969. Oriole 35:1-9.

Fink, L. C. and T. W. French. 1971. Birds in downtown Atlanta – Fall 1970. Oriole 35:13-20.

Fisher, J. D., S. C. Schneider, A. A. Ahlers, and J. R. Miller. 2015. Categorizing wildlife responses to urbanization and conservation implications of terminology: Terminology and urban conservation. Conservation Biology 29:1246-1248.

FLAP Canada. 2018. An analysis of collision mitigation effectiveness pre- and post-installation of bird collision deterrents at four Toronto buildings. [Unpublished Report].

Forbush, E. H. 1929. Birds of Massachusetts and other New England States. Part III. Land birds from sparrows to thrushes. Massachusetts Department of Agriculture.

Fraser, D. 2014. Birds and windows: an animal welfare problem? AnimalSense 15:23.

Gang, J. 2011. Reveal: Studio Gang Architects. Princeton Architectural Press, New York, USA.

Ganier, A. F. 1963. A Kentucky bird-fall in 1962. Migrant 34:34-35.

Garces, A., I. Pires, F. Pacheco, L. Sanches Fernandes, V. Soeiro, S. Loio, J. Prada, R. Cortes, and F. Queiroga. 2020. Natural and anthropogenic causes of mortality in wild birds in a wildlife rehabilitation centre in Northern Portugal: a ten-year study. Bird Study, doi.10.1080/00063657.2020.1726874.

Gaston, K. J. and T. M. Blackburn. 1997. How many birds are there? Biodiversity and Conservation 6:615-625.

Gaston, K. J. and R. A. Fuller. 2008. Commonness, population depletion and conservation biology. Trends in Ecology and Evolution 23:14-19. doi:10.1016/j.tree.2007.11.001.

Gaston, K. J., T. W. Davis, J. Bennie, and J. Hopkins. 2012. Reducing the ecological consequences of night-time light pollution: options and developments. Journal of Applied Ecology 49:1256-1266.

Gauthreaux, S., Jr., and C. Belser. 2006. Effects of artificial night lighting on migrating birds, in C. Rich and T. Longcore, Editors, Ecological consequences of Artificial Night Lighting, Island Press, Covelo, California and Washington, D. C., 67-93.

Gehring, J., P. Kerlinger, and A. M. Manville, II. 2009. Communication towers, lights, and birds: successful methods of reducing the frequency of avian collisions. Ecological Applications 19:505-514.

Gelb, Y. and N. Delacretaz. 2006. Avian window strike mortality at an urban office building. The Kingbird 56: 190-198.

Gelb, Y. and N. Delacretaz. 2009. Windows and vegetation: primary factors in Manhattan bird collisions. Northeastern Naturalist 16:455-470 DOI 10.1656/045.016.n312.

George, W. G. 1968. Check list of birds of southern Illinois. Southern Illinois University at Carbondale, Carbondale, Illinois, USA [Mimeographed Unpublished Manuscript].

Gibson, E. J. and R. D. Walk. 1960. The "visual cliff." Scientific American 202:64-71.

Ghim, M. M., and W. Hodos. 2006. Spatial contrast sensitivity of birds. Journal of Comparative Physiology A 192:523-534. DOI 10.1007/s00359-005-0090-5.

Gill, F. B., and R. O. Prum. 2019. Ornithology, 4rd Edition. W. H. Freeman and Company, New York, New York, USA.

Giller, F. 1960. Eine moderne "Vogelfalle." Ornithologische Mitteilungen 12:152-153.

Goldsmith, T. H. 2006. What birds see. Scientific American 295:68-75.

Goller, B., B. F. Blackwell, T. L. DeVault, P. E. Baumhardt, and E. Fernandez-Juricic. 2018. Assessing bird avoidance of high-contrast lights using a choice test approach: Implications for reducing human-induced avian mortality. PeerJ doi:10.7717/peerj.5404.

Gomez-Martinez, M. A., D. Klem, Jr., O. Rojas-Soto, F. Gonzalez-Garcia, and I. MacGregor-Fors. 2019. Window strikes: bird collisions in a Neotropical green city. Urban Ecosystems. https://doi.org/10.1007/s11252-019-00858-6.

Gomez-Moreno, V. del Carmen, J. R. Herrera-Herrera, S. Nino-Maldonado. 2018. Colision de aves en ventanas del Centro Universitario Victoria, Tamaulipas, Mexico. [Bird collisions in windows of Centro Universitario Victoria, Tamaulipas, Mexico.] Huitzil, Revista Mexicana de Ornitologia 19:227-236. https://doi.org/10.28947/hrmo.2018.19.2.347.

Graber, R. R. 1968. Nocturnal migration in Illinois: different points of view. Wilson Bulletin 80:36-71.

Graham, D. L. 1997. Spider webs and windows as potentially important sources of hummingbird mortality. Journal of Field Ornithology 68(1): 98-101.

Habberfield, M. W. and C. C. St. Clair. 2016. Ultraviolet lights do not deter songbirds at feeders. Journal of Ornithology 157:239-248. DOI 10.1007/s10336-015-1272-8.

Hadidian, J. 2007. Wild neighbors: the humane approach to living with wildlife, 2nd Edition. Humane Society of the United States, Washington D.C., USA.

Hager, S. B. 2009. Human-related threats to urban raptors. Journal of Raptor Research 43: 210-226.

Hager, S. B., and B. J. Cosentino. 2014. Surveying for bird carcasses resulting from window collisions: standardized protocol. PeerJ PrePrints 2:e406v401 DOI: 10.7287/peerj.preprints.406v1.

Hager, S. B., B. J. Cosentino, K. J. McKay. 2012. Scavenging affects persistence of avian carcasses resulting from window collisions in an urban landscape. Journal of Field Ornithology 83:203-211 DOI 10.1111/jofo.2012.83.issue-2.

Hager, S. B., B. J. Cosentno, M. A. Aguilar-Gomez, M. L. Anderson, M. Bakermans, T. J. Boves, B. Brandes, M. W. Butler, E. M. Butler, N. L. Cagle, R. Calderon-Parra, A. P. Capparella, A. Chen, K. Cipollini, A. A. T. Conkey, T. A. Contreras, R. I. Cooper, C. E. Corbin, R. L. Curry, J. J. Dosch, M. G. Drew, K. Dyson, C. Foster, C. D. Francis, E. Fraser, R. Furbush, N. D. G. Hagemeyer, K. N. Hopfensperger, D. Klem, Jr., E. Lago, A. Lahey, K. Lamp, G. Lewis, S. R. Loss, C. S. Machtans, J. Madosky, T. J. Maness, K. J. McKay, S. B. Menke, K. E. Muma, N. Ocampo-Penuela, T. J. O'Connell, R. Ortega-Alvarez, A. L. Pitt, A. L. Puga-Caballero, J. E. Quinn, C. W. Varian-Ramos, C. S. Riding, A. M. Roth, P. G. Saenger, R. T. Schmitz, J. Schnurr, M. Simmons, A. D. Smith, D. R. Sokoloski, J. Vigliotti, E. L. Walters, L. A. Walters, J. T. Weir, K. Winnett-Murray, J. C. Withey, and I. Zuria. 2017. Continent-wide analysis of how urbanization affects bird-window collision mortality in North America. Biological Conservation 212:209-215. http://dx.doi.org/10.1016/j.biocon.2017.06.014

Hager, S. B., B. J. Cosentino, K. J. McKay, C. Monson, W. Zuurdeeg, and B. Blevins. 2013. Window area and development drive spatial variation in bird-window collisions in an urban landscape. PLoS One 8, e53371.

Hager, S. B., and M. E. Craig. 2014. Bird-window collisions in the summer breeding season. PeerJ 2:e460 DOI 10.7717/peerj.460.

Hager, S. B., H. Trudell, K. J. McKay, S. M. Crandall, and L. Mayer. 2008. Bird density and mortality at windows. Wilson Journal of Ornithology 120: 550-564.

Hall, G. A. 1972. Fall migration – Appalachian region. American Birds 26:62-66.

Hall, G. A. 1974. Fall migration – Appalachian region. American Birds 28:52.

Harden, J. 2002. An overview of anthropogenic causes of avian mortality. Journal of Wildlife Rehabilitation 25:4-11.

Harwin, R. M. 1978. The lesser window-bashing kingfisher. The Honeyguide 93:19-20.

Hastad, O. and A. Odeen. 2014. A vision physiological estimation of ultraviolet window marking visibility to birds. PeerJ 2:e621;DOI 10.7717/peerj.621.

Haupt, H. 2011a. Auf dem Weg zu einem neuen Mythos? Warum UV-Glas zur Vermeidung von Vogelschlag noch nicht empfohlen warden kann. Berichte zum Vogelschutz 47/48: 143-160.

Haupt, H. 2011b. Massen-Irritation ziehender Singvogel durch StraBenbeleuchtung. Berichte zum Vogelschutz 47/48: 161-165.

Haupt, V. H., and U. Schillemeit. 2011. Skybeamer und Gebaudeanstrahlungen bringen Zugvogel vom Kurs ab: Neue Untersuchungen und eine rechtliche Bewertung dieser Lichtanlagen. Naturschutz und Landschaftsplaung 43:165-170.

Hausberger, M., A. Boigne, C. Lesimple, L. Belin, and L. Henry. 2018. Wide-eyed glare scares raptors: From laboratory evidence to applied management. PLoS ONE 13(10):e0204802. https://doi.org/10.1371/journal.pone.0204802.

Hausmann, F., K. E. Arnold, N. J. Marshall, and I. P. F. Owens. 2003. Ultraviolet signals in birds are special. Proceedings of the Royal Society of London B 270:61-67.

Henderson, C. L. 1987. Landscaping for wildlife. Department of Natural Resources, St. Paul, Minnesota, USA.

Herbert, A. D. 1970. Spatial disorientation in birds. Wilson Bulletin 82:400-419.

Hodos, W. 2012. What birds see and what they don't: Luminance, contrast, and spatial and temporal resolution. Chapter 1 in Lazareva, O. F., T. Shimizu, and E. A. Wasserman. How animals see the world Comparative behavior, biology, and evolution of vision. Oxford University Press, Oxford, U.K., 5-24.

Hoeck, P., and B. Rideout. 2016. Testing of a promising UV window film to avoid bird window collisions. Project Report, San Diego Zoo Global. [Unpublished Manuscript].

Homayoun, T. Z., and R. B. Blair. 2016. Value of park reserves to migrating and breeding landbirds in an urban important bird area. Urban Ecosystems 19:1579-1596.

Hong, D. 2006. Preventing bird-window collisions: adding eyespots to falcon silhouettes. EAP Tropical Biology Program, Costa Rica, University of California, Riverside [Unpublished Manuscript].

Horton, J. 2013. Surefire ways to make windows friendly to birds. BirdWatching April:32-37.

Horton, K. G., C. Nilsson, B. M. Van Doren, F. A. La Sorte, A. M. Dokter, and A. Farnsworth. 2019. Bright lights in the big cities: Migratory birds' exposure to artificial light. Frontiers in Ecology and the Environment. doi:10.1002/fee.2029.

Horvath, G., G. Kriska, P. Malik, and B. Robertson. 2009. Polarized light pollution: a new kind of ecological photopollution. Frontiers in Ecology and Environment 7:317-325.

Horvath, G., G. Kriska, and B. Robertson. 2014. Anthropogenic polarization and polarized light pollution inducing polarized ecological traps. In Polarized light and polarization vision in animal sciences, 443-513. Springer, Berlin, Heidelberg, Germany.

Houppert, K. 2014. Lights out. Washington Post Magazine. 16 March:18-25.

Huggins, B. and S. Schlacke. 2019. Rechtliche Anforderungen und Gestaltungsmoglichkeiten. [Protection of species from glass and light: Legal requriements and design options.] Springer. ISBN 978-3-662-58257-2.

Hunt, S., A. T. D. Bennett, I. C. Cuthill, and R. Griffith. 1998. Blue tits are ultraviolet tits. Proceedings of Royal Society 265: 451-455.

Hutchins, M., P. O. Marra, E. Diebold, M. D. Kreger, C. Sheppard, S. Hallager, and C. Lynch. 2018. The evolving role of zoological parks and aquariums in migratory bird conservation. Zoo Biology 1-9. https://doi.org/10.1002/zoo.21438.

Ingrassia, N. 2016. Does sound help prevent birds from flying into objects? Master of Science Thesis, College of William and Mary, Williamsburg, Virginia, USA.

Irwin, M. P. S. 1978. The life and death of the pygmy kingfisher (*Ispidina picta*) in Rhodesia. The Honeyguide 93:29-33.

Jakle, J. 2001. City lights: illuminating the American night. Johns Hopkins University Press, Baltimore, Maryland, USA.

Jenkins, A. R., J. J. Smallie, M. Diamond. 2010. Avian collisions with power lines: a global review of causes and mitigation with a South African perspective. Bird Conservation International 20:263-278.

Johnston, D. W., and T. P. Haines. 1957. Analysis of mass bird mortality in October, 1954. Auk 74:447-458.

Johnson, R. E. and G. E. Hudson. 1976. Bird mortality at a glassed-in walkway in Washington State. Western Birds 7:99-107.

Jones, H. L. 2005. Central America. North American Birds 59: 162-165.

Jones, J. and C. M. Francis. 2003. The effects of light characteristics on avian mortality at light houses. Journal of Avian Biology 34:328-333.

Kahle, L. Q., M. E. Flannery, and J. P. Dumbacher. 2016. Bird-window collisions at a west-coast urban park museum: analyses of bird biology and window attributes from Golden Gate Park, San Francisco. PLoS ONE 11(1):e0144600. DOI:10.1371/journal.pone.0144600.

Keil, W. 1964. Der Glaserne Tod. Gefahrdung der Vogelwelt durch Glaswande. Vogel-Kosmos 1:184-186.

Kemper, C. 1996. A study of bird mortality at a west-central Wisconsin TV tower from 1957-1995. Passenger Pigeon 58:219-235.

Kenney, D. T. 2013. Aesthetic danger: how the human need for light and spacious views kills birds and what we can (and should) do to fix this invisible hazard. Journal of Animal and Natural Resource Law 11:137-159.

Kensek, K., Y. Ding, and T. Longcore. 2016. Green building and biodiversity: Facilitating bird friendly design with building information models. Journal of Green Building 11:116-130. doi:10.3992/jgb.11.2.116.1.

Kerlinger, P., J. L. Gehring, W. P. Erickson, R. Curry, A. Jain, and J. Guarnaccia. 2010. Night migrant fatalities and obstruction lighting at wind turbines in North America. The Wilson Journal of Ornithology 122:744-754. doi:10.1676/06-075.1.

Keyes, T., and L. Sexton. 2014. Characteristics of bird strikes at Atlanta's commercial buildings during late summer and fall migration, 2005. The Oriole:79.

Kiltie, R. A. 2000. Scaling of visual acuity with body size in mammals and birds. Functional Ecology 14:226-234.

King, J. R. and W. J. Bock. 1978. Workshop on a national plan for ornithology final report. [Mimeographed Unpublished Manuscript].

Kirby, J. S., A. J. Stattersfield, S. H. Butchart, M. I. Evans, R. F. Grimmett, V. R. Jones, J. O'Sullivan, G. M. Tucker, and I. Newton. 2008. Key conservation issues for migratory land- and waterbird species on the world's major flyways. Bird Conservation International 18:49-S73.

Klem, D. Jr. 1979. Biology of collisions between birds and windows. Ph.D. Dissertation, Southern Illinois University at Carbondale, Illinois, xiii+256.

Klem, D. Jr. 1981. Avian predators hunting birds near windows. Proceedings of the Pennsylvania Academy of Science 55:90-92.

Klem, D. Jr. 1985. Window pains. The Living Bird Quarterly 4: 21.

Klem, D. Jr. 1989. Bird-window collisions. Wilson Bulletin 101(4): 606-620.

Klem, D. Jr. 1990a. Bird injuries, cause of death, and recuperation from collisions with windows. Journal of Field Ornithology 61(1): 115-119.

Klem, D. Jr. 1990b. Collisions between birds and windows: mortality and prevention. Journal of Field Ornithology 61(1): 120-128.

Klem, D. Jr. 1991. Glass and bird kills: an overview and suggested planning and design methods of preventing a fatal hazard. Pages 99-104 *in* L. W. Adams and D. L. Leedy, editors.Wildlife Conservation in Metropolitan Environments NIUW Symposium Series 2, National Institute for Urban Wildlife, Maryland, USA.

Klem, D. Jr. 1992. An invisible killer. Bird Watcher's Digest 14:80-90.

Klem, D. Jr. 2004. Glass – an unintended but catastrophic hazard for birds. Animal Welfare Institute Quarterly 53:4-5.

Klem, D., Jr. 2006. Glass: a deadly conservation issue for birds. Bird Observer 34:73-81.

Klem, D., Jr. 2007. Windows: an unintended fatal hazard for birds. Pages 7-12 *in* Connecticut State of the Birds 2007. Connecticut Audubon Society, Fairfield, Connecticut, USA.

Klem, D., Jr. 2008. Bird window-kills and window-kill records. Winging It 20:21.

Klem, D., Jr. 2009a. Avian mortality at windows: the second largest human source of bird mortality on earth. In Proceedings of the Fourth International Partners in Flight Conference: Tundra to Tropics, McAllen, TX, 244-251.

Klem, D., Jr. 2009b. Preventing bird-window collisions. Wilson Journal of Ornithology 121:314-321.

Klem, D., Jr. 2010. Conservation issues: Sheet glass as a principal human-associated avian mortality factor. In S. K. Majumdar, T. L. Master, R. M. Ross, R. Mulvihill, M. Brittingham, and J. Huffman [eds.] Avian ecology and conservation: A Pennsylvania focus with national implications, Pennsylvania Academy of Science, Harrisburg, PA.

Klem, D., Jr. 2012a. Method and apparatus for preventing birds from colliding with or striking flat clear and tinted glass and plastic surfaces. United States Patent, Patent Number US 8,114,503 B2.

Klem, D., Jr. 2012b. Method and apparatus for preventing birds from colliding with or striking flat clear and tinted glass and plastic surfaces. United States Patent, Patent Number US 2012/0113519 A1.

Klem, D., Jr. 2014. Landscape, legal, and biodiversity threats that windows pose to birds: a review of an important conservation issue. Land 3:351-361; doi:10.3390/land3010351.

Klem, D., Jr. 2015. Bird-window collisions: a critical animal welfare and conservation issue. Journal of Applied Animal Welfare Science, 18:Sup1,S11-S17, DOI:10.1080/10 888705.2015.1075832.

Klem, D., Jr. 2018. Erratum for Klem (2015). Journal of Applied Animal Welfare Science, 21:101, DOI:10.1080/10888705.2017.1396838.

Klem, D., Jr. 2019. A national standard for bird-friendly building design. Construction Canada 61:40-44.

Klem, D., Jr., C. R. Brancato, J. F. Catalano, and F. L. Kuzmin. 1982. Gross morphology and general histology of the esophagus, ingluvies and proventriculus of the House Sparrow (*Passer domesticus*). Proceedings of the Pennsylvania Academy of Science 56:141-146.

Klem, D., Jr., S. A. Finn, and J. H. Nave, Jr. 1983. Gross morphology and general histology of the ventriculus, intestinum, caeca and cloaca of the House Sparrow (*Passer domesticus*). Proceedings of the Pennsylvania Academy of Science 57:27-32.

Klem, D., Jr., M. A. Parker, W. L. Sprague, S. A. Tafuri, C. J. Veltri, and M. J. Walker. 1984. Gross morphology and general histology of the alimentary tract of the American Robin (*Turdus migratorius*). Proceedings of the Pennsylvania Academy of Science 58:151-158.

Klem, D., Jr., D. C. Keck, K. L. Marty, A. J. Miller Ball, E. E. Niciu, and C. T. Platt. 2004. Effects of window angling, feeder placement, and scavengers on avian mortality at plate glass. Wilson Bulletin 116:69-73.

Klem, D., Jr., C. J. Farmer, N. Delacretaz, Y. Gelb, and P. G. Saenger. 2009. Architectural and landscape risk factors associated with bird-glass collisions in an urban environment. Wilson Journal of Ornithology 121:126-134.

Klem, D., Jr., K. L. DeGroot, E. A. Krebs, K. T. Fort, S. B. Elbin, and A. Prince. 2011. A second critique of 'Collision mortality has no discernible effect on population trends of North American Birds'. PLoS One 6, e24708.

Klem, D., Jr. and P. G. Saenger. 2013. Evaluating the effectiveness of select visual signals to prevent bird-window collisions. Wilson Journal of Ornithology 125:406-411.

Klemens, S., K. Baganz, and R. Altenkamp. 2017. Vogelschlag an Glasflachen von Tiergehegen. Tiergarten 4:36-51.

Kolbert, E. 2014. The sixth extinction: an unnatural history. Picador, Henry Holt and Company, New York, USA.

Konig, C. 1963. Glaswande als Gefahren fur die Vogelwelt. Deutsche Sektion des Internationalem Rates für Vogelschutz, Bericht 2:53-55.

Korner-Nievergelt, F., O. Behr., R. Brinkmann, M. A. Etterson, M. M. P. Huso, D. Dalthorp, P. Korner-Nievergelt, T. Roth, and I. Niermann. 2015. Mortality estimation from carcass searches using the R-package carcass – A tutorial. Wildlife Biology 21:30-43.

Korner-Nievergelt, F., R. Brinkmann, I. Niermann, and O. Behr. 2013. A new method to determine bird and bat fatality at wind energy turbines from carcass searches. Wildlife Biology 17:350-363.

Kramer, Q. 1948. Bird tragedy in a fog. Cassinia 37:21-22.

Kreithen, M. L., and T. Eisner. 1978. Detection of ultraviolet light by the homing pigeon. Nature 272:347-348.

Kriska, G., P. Malik, I. Szivak, and G. Horvath. 2008. Glass buildings on river banks as "polarized light traps" for mass-swarming polarotactic caddis flies. Naturwissenschaften 95:461-467. doi:10.1007/s00114-008-0345-4.

Kudelka, M. 1994. See no evil: Glass is a clear and present danger to migrating birds. Nature Canada 23:38-41.

Kummer, J. A., and E. M. Bayne. 2015. Bird feeders and their effects on bird-window collisions at residential houses. Avian Conservation and Ecology 10:6.

Kummer, J. A., E. M. Bayne, and C. S. Machtans. 2016a. Use of citizen science to identify factors affecting bird-window collision risk at houses. Condor 118:624-639. DOI: http://dx.doi.org/10.1650/CONDOR-16-26.1.

Kummer, J. A., E. M. Bayne, and C. S. Machtans. 2016b. Comparing the results of recall surveys and standardized searches in understanding bird-window collisions at residential houses. Avian Conservation and Ecology 11:4.

Kummer, J. A., C. J. Nordell, T. M. Berry, C. V. Collins, C. R. L. Tse, and E. M. Bayne. 2016. Use of bird carcass removals by urban scavengers to adjust bird-window collision estimates. Avian Conservation and Ecology 11:12. Online at http://www.ace-eco.org/vol11/iss2/art12/.

Kuntz, T. H., S. A. Gauthreaux, Jr., N. I. Hristov, J. W. Horn, G. Jones, E. K. V. Kalko, R. P. Larkin, G. F. McCracken, S. M. Swarty, R. B. Srygley. 2008. Aeroecology: Probing and modeling the aerosphere. Integrative and Comparative Biology 48:1-11. https://doi.org/10.1093/icb/icn037.

Kuntzman, G. 2001. The glass trap. Metropolis 20:36.

Labedz, T. E. 1997. Windows of death: a look at bird strikes. University of Nebraska State Museum, Museum Notes 95:1-4.

Lack, D. 1954. The natural regulation of animal numbers. Oxford University Press, Ely House, London, U.K.

Lack, D. 1960. The influence of weather on passerine migration. A review. Auk 77:171-209.

La Sorte, F. A., D. Fink, J. J. Buler, A. Farnsworth, and S. A. Cabrera-Cruz. 2017. Seasonal associations with urban light pollution for nocturnally migrating bird populations. Glob Chang Biol 23:4609-4619. doi:10.111/gcb.13792.

Lambertucci, S. A., E. L. C. Shepard, and R. Wilson. 2015. Human-wildlife conflicts in a crowded airspace. Science 348:502-504.

Land, M. F. 1999. Motion and vision: why animals move their eyes. Journal of Comparative Physiology A 185:341-352.

Land, M. F., and D-E. Nilsson. 2010. Animal eyes. Oxford University Press, New York, New York, USA.

Lang, R., T. W. Sanchez, and A. C. Oner. 2009. Beyond edge city: office geography in the new metropolis. Urban Geography 30:726-755.

Langridge, H. P. 1960. Warbler kill in the Palm Beaches. Florida Naturalist 33:226-227.

Lao, S., B. A. Robertson, A. W. Anderson, R. B. Blair, J. W. Eckles, R. J. Turner, and S. R. Loss. 2020. The influence of artificial light at night and polarized light on bird-building collisions. Biological Conservation 241. https://doi.org/10.1016/j.biocon.219.108358.

Lapedes, D. N. Editor. 1978. McGraw-Hill dictionary of physics and mathematics. McGraw-Hill Book Company, New York, USA.

Leibach, J. 2008. Pain in the glass. Audubon 110:84-87.

Leonard, P. 2014. Glass action for birds. Living Bird 33:26-31.

Ley, H. W. 2006. Experimental examination of the perceptibility of patented bird-protecting glass to a sample of Central European perching birds. Max Planck Institute for Ornithology. [Unpublished Report].

Ley, H-W, and W. Fiedler. 2010. Suitability of bird-safe glass panes for transparent noise barriers. Max Planck Institute for Ornithology [Unpublished Manuscript].

Lights Out Columbus. 2012. Lights out Columbus monitoring program spring 2012 report. Online at http://www.lightsoutcolumbus.org/.

Lin, H., I. G. Ros, and A. A. Biewener. 2014. Through the eyes of a bird: modeling visually guided obstacle flight. Journal of the Royal Society Interface11, 96:20140239.

Lincoln. F. C. 1931. Some causes of mortality among birds. Auk 48:538-546.

Liu, H., and X. Yanchun. 2017. Bird collision with building glass outer wall caused by landscape structure: A case study, College of Wildlife Resources, Northeast Forestry University. Chinese Journal of Wildlife 2014-02.

Lohrl, von H. 1962a. Vogelvernichtung durch moderne Glaswande. Kosmos 5:191-194.

Lohrl, von H. 1962b. Tatigkeitsbericht der Staatlichen Vogelschutzwarte Ludwigsburg in Veroffentl. d. Landesst. fur Naturschutz und Landschaflspflege Baden-Wuttemberg 30:252.

Long, L. and D. DeAre. 1982. Repopulating the countryside: a 1980 census trend. Science 217:1111-1116.

Longcore, T., and C. Rich. 2004. Ecological light pollution. Frontiers in Ecology and the Environment 2:191. doi:10.2307/3868314.

Longcore, T., C. Rich, P. Mineau, B. MacDonald, D. G. Bert, L. M. Sullivan, E. Mutrie, S. A. Gauthreaux, Jr., M. L. Avery, R. L. Crawford, A. M. Manville II, E. R. Travis, and d. Drake. 2012. An estimate of avian mortality at communication towers in the United States and Canada. PLoS ONE 7(4):e34025 DOI 10.1371/journal.pone.0034025.

Longcore, T., C. Rich, P. Mineau, B. MacDonald, D. G. Bert, M. L. Sullivan, and D. Drake. 2013. Avian mortality at communication towers in North America: which species, how many, and where? Biological Conservation 158:410-419.

Longcore, T., and P. A. Smith. 2013. On avian mortality associated with human activities. Avian Conservation Ecology 8:1.

Loss, S. R., M. O. Ruiz, and J. D. Brawn. 2009. Relationships between avian diversity, neighborhood age, income, and environmental characteristics of an urban landscape. Biological Conservation 142:2578-2585.

Loss, S. R., S. S. Loss, T. Will, and P. P. Marra. 2014. Best practices for data collection in studies of bird-window collisions. Online at http://www.dropbox.com/s/q4m8zboobvk7zjt/Best%20Practices%20for%20Window%20Data%20FINAL%20eVersion.docx.

Loss, S. R., S. S. Loss, T. Will, and P. P. Marra. 2015. Linking place-based citizen science with large-scale conservation research: a case study of bird-building collisions and the role of professional scientists. Biological Conservation 184:439-445.

Loss, S. R., T. Will, and P. P. Marra. 2012a. Direct human-caused mortality of birds: improving quantification of magnitude and assessment of population impact. Frontiers in Ecology and the Environment 10: 357-364.

Loss, S. R., T. Will, and P. P. Marra. 2012b. The impact of free-ranging domestic cats on wildlife of the United States. Nature Communications 4:1396.

Loss, S. R., T. Will, S. S. Loss, and P. P. Marra. 2014. Bird-building collisions in the United States: estimates of annual mortality and species vulnerability. Condor: Ornithological Applications 116:8-23.

Loss, S. R., T. Will, and P. P. Marra. 2015. Direct mortality of birds from anthropogenic causes. Annual Review of Ecology, Evolution, and Systematics 46:99-120, DOI: 10.1146/annurev-ecolsys-112414-054133.

Loss, S. R., and P. P. Marra. 2017. Population impacts of free-ranging domestic cats on mainland vertebrates. Frontiers in Ecology and the Environment 15:502-509, DOI: 10.1002/fee.1633.

Loss, S. R., S. Lao, J. W. Eckles, A. W. Anderson, R. B. Blair, and R. J. Turner. 2019. Factors influencing bird-building collisions in the downtown area of a major North American city. PLoS ONE 14(11):e0224164. https://doi.org/10.1371/journal.pone.0224164.

Loviglio, J. 2004. Glass windows an 'Indiscriminate' Bird Killer. http://www.cnn.com/2004/TECH/science/02/03/birds.clear.danger.ap/.

Low, B. W., D. L. Yong, D. Tan, A. Owyong, and A. Chia. 2017. Migratory bird collisions with man-made structures in South-East Asia: a case study from Singapore. BirdingASIA 27:107-111.

Low, T. 2014. Where song began: Australia's birds and how they changed the world. Penguin Random House Australia.

Lowry, S. 1992. Injuries from domestic glazing. British Medical Journal 304:332.

Lowther, P. E. 1995. Ornithology at the Field Museum. Pages 145-161 in W. E. Davis, Jr. and J. A. Jackson, editors. Contributions to the history of North American Ornithology. Memoirs of the Nuttall Ornithological Club, No. 12, Cambridge, Massachusetts, USA.

Lozano, R., K. Ceulemans, M. Alonso-Almeida, D. Huisingh, F. J. Lozano, T. Waas, W. Lambrechts, R. Lukman, and J. Huge. 2015. A review of commitment and implementation of sustainable development in higher education: Results from a worldwide survey. Journal of Cleaner Production 108:1-18. DOI:10.1016/j.jclepro.2014.09.048.

Lyytinen, A., P. M. Bakefield, and J. Mappes. 2003. Significance of butterfly eyespots as an antipredator device in ground-based and aerial attacks. Oikos 100:373-379.

MacGregor-Fors, I., and J. E. Schondube. 2011. Gray vs. green urbanization: relative importance of urban features for urban bird communities. Basic Applied Ecology 12:372-381.

Machtans, C. S., C. H. R. Wedeles, and E. M. Bayne. 2013. A first estimate for Canada of the number of birds killed by colliding with building windows. Avian Conservation Ecology 8(2):6, doi:10.5751/ACE-00568-080206.

Machtans, C. S. and W. E. Thogmartin. 2014. Understanding the value of imperfect science from national estimates of bird mortality from window collisions. Condor: Ornithological Applications 116: 3-7, DOI:10.1650/CONDOR-13-134.1.

Mahjoub, G., M. K. Hinders, and J. P. Swaddle. 2015. Using a "sonic net" to deter pest bird species: Excluding European starlings from food sources by disrupting their acoustic communication. Wildlife Society Bulletin 39:326-333. doi:10.1002/wsb.529.

Maier, E. J., and J. K. Bowmaker. 1993. Colour vision in the passeriform bird, Leiothrix lutea: Correlation of visual pigment absorbance and oil droplet transmission with spectral sensitivity. Journal of Comparative Physiology A 172:295-301. doi:10.1007/bf00216611.

Maitra, A. and K. Han. 1989. Architectural glass injuries: a case for effective prevention. Archives of Emergency Medicine 6:169-171.

Malakoff, D. 2004. Clear & present danger. Audubon 106:65-68.

Mannan, R. W., R. J. Steidl, and C. W. Boal. 2008. Identifying habitat sinks: a case study of Cooper's Hawks in an urban environment. Urban Ecosystems 11:141-148.

Manville, II, A. 2009. Towers, turbines, powerlines, and buildings – steps being taken by the U.S. Fish and Wildlife Service to avoid or minimize take of migratory birds at these structures. Proceedings of the Fourth International Partners in Flight Conference: Tundra to Tropics, Partners in Flight, McAllen, Texas, USA 262-272.

Markandya, A., T. Taylor, A. Longoc, N. N. Murty, S. Murtyd, and K. Dhavalad. 2008. Counting the cost of vulture decline – An appraisal of the human health and other benefits of vultures in India. Ecological Economics 67:194-204. https://doi.org/10.1016/j.ecolecon.2008.04.020.

Martin, G. R. 2007. Visual fields and their functions in birds. Journal of Ornithology 148:547-562.

Martin, G. R. 2009. What is binocular vision for? A bird's eye view. Journal of Vision 9(11). https://doi.org/10.1167/9.11.14, 14, 1-19. http://journalofvision.org/9/11/14/.

Martin, G. R. 2011. Understanding bird collisions with man-made objects: a sensory ecology approach. Ibis 153:239-254.

Martin, G. R. 2012. Through birds' eyes: insights into avian sensory ecology. Journal of Ornithology 153(Suppl 1):S23-S48. DOI 10.1007/s10336-011-0771-5.

Martin, G. R. 2017. The sensory ecology of birds. Oxford University Press, Oxford, UK.

Matthews, S. N., and P. G. Rodewald. 2010. Movement behavior of a forest songbird in an urbanized landscape: the relative importance of patch-level effects and body condition during migratory stopover. Landscape Ecology 25:955-965.

Marzluff, J. M. 2001. Worldwide urbanization and its effects on birds. In: Avian Ecology and Conservation in an Urbanizing World 19-47. https://doi.org/10.1007/978-1-4615-1531-9_2.

Marzluff, J. M., R. Bowman, and r. Donnelly, Editors. 2001. Avian ecology and conservation in an urbanizing world. Kluwer Academic Publishers, Boston, Massachusetts, USA.

May, R., J. Astrom, O. Hamre, and E. L. Dahl. 2017. Do birds in flight respond to (ultra)violet lighting? Avian Research 8:33. https://doi.org/10.1186/s406657-017-0092-3.

McCommons, J. 2007. It is as clear as day. Wildlife Conservation 110:20-25.

McKinney, M. L. 2002. Urbanization, biodiversity, and conservation: The impacts of urbanization on native species are poorly studied, but educating a highly urbanized human population about these impacts can greatly improve species conservation in all ecosystems. BioScience 52:883-890. https://doi.org/10.1641/0006-3568(2002)052[0883:UBAC]2.0.CO;2.

McKinney, M. L. 2006. Urbanization as a major cause of biotic homogenization. Biological Conservation 127:247-260.

McLaren, J. D., J. J. Buler, T. Schreckengost, J. A. Smolinsky, M. Boone, E. E. van Loon, D. K. Dawson, and E. L. Walters. 2018. Artificial light at night confounds broad-scale habitat use by migrating birds. Ecological Letters 21:356-364.

Melles, S., S. Glenn, and K. Martin. 2003. Urban bird diversity and landscape complexity: species-environment associations along a multiscale habitat gradient. Conservation Ecology 7:5.

Menacho-Odio, R. M. 2015. Colision de aves contra ventanas en Costa Rica: conociendo el problema a partir de datos de museos, ciencia ciudadana y el aporte de biologos. Zeledonia 19:10-21.

Menacho-Odio, R. M. 2018. Colision de aves con ventanas: problema, prevencion, mitigacion y tendencias de investigacion [Bird-window collisions: Problem, prevention, mitigation and research directions.]. Zeledonia 22:59-76.

Menacho-Odio, R. M. 2018. Local perceptions, attitudes, beliefs, and practices toward bird-window collisions in Monteverde, Costa Rica. UNED Research Journal 10:33-40.

Menacho-Odio, R. M., M. Garro-Cruz, and J. E. Arevalo. 2019. Ecology, endemism, and conservation status of birds that collide with glass windows in Monteverde, Costa Rica. Revista de Biologia Tropical 67:S326-S345.

Mercer, B. 1905. A municipal bird trap. American Ornithology 5:53-55.

Merkel, F. R. and K. L. Johansen. 2011. Light induced bird-strikes on vessels in South West Greenland. Marine Pollution Bulletin 62:2330-2336.

Milius, S. 2013. Collision course: scientists struggle to make windows safer for birds. Science News 184:20-25.

Milius, S. 2014. Searching for bird-safe windows. National Wildlife 52:14-16.

Miller, A. H. 1946. A method of determining the age of live passerine birds. Bird Banding 17:33-35.

Mills, A. M. 2016. Banding data reveal bias in age-class sampling of songbirds during spring migration. Journal of Field Ornithology 87:323-336.

Mitrus, C., and A. Zbyryt. 2017. Reducing avian mortality from noise barrier collisions along an urban roadway. Urban Ecosystems. doi.org/10.1007/s11252-017-0717-7.

Monge-Najera, J. Llosa, and Z. B. Llosa. 2018. A new, cheap method to reduce bird mortality from window collisions. UNED Research Journal 10:83-84.

Morris, S. R. A. R. Clark, L. H. Bhatti, and H. J. L. Glasgow. 2003. Television tower mortality of migrant birds in western New York and Youngstown, Ohio. Northeast Naturalist 10:67-77.

Morrison, M. L., A. D. Rodewald, G. Voelker, M. R. Colon, and J. F. Prather. 2018. Ornithology, Foundation, Analysis, and Application. Johns Hopkins University Press, Baltimore, Maryland, USA.

Morzer Bruijns, M. D., and L. J. Stwerka. 1961. Het doodvliegenvan vogels tegen ramen. De Levende Natuur 64:253-257.

Moxley, A. 2008. Flight risk. All Animals 10:23-24.

Myers, N., R. A. Mittermeier, C. G. Mittermeier, G. A. Da Fonseca, and J. Kent. 2000. Biodiversity hotspots for conservation priorities. Nature 403:853-858.

National Audubon Society. 2007. The 2007 Audubon WatchList. Available from http://web1.audubon.org/science/watchlist/browseWatchlist.php (accessed November 2007).

National Audubon Society. 2013. Important Bird Areas in the U.S. National Audubon Society, New York, USA. Online at www.audubon.org/bird/iba.

New York City Audubon. 2007. Bird-safe Building Guidelines, New York City Audubon, 71 W. 23rd Street, New York, New York, USA.

Newman, R. J. and G. H. Lowery, Jr. 1959. The changing seasons: spring migration 1959. Audubon Field Notes 13:350.

Newton, I., I. Wyllie, and L. Dale. 1999. Trends in the numbers and mortality patterns of Sparrowhawks (*Accipiter nisus*) and Kestrels (*Falco tinnunculus*) in Britain, as revealed by carcass analyses. Journal of Zoology 248:139-147.

Nichols, K. S. 2019. Birds and buildings: Bird-window collisions in the urban landscape. Ph.D. Dissertation, University of Minnesota. https://conservancy.umn.edu/handle/11299/200288.

Nichols, K. S., T. Homayoun, J. Eckles, R. B. Blair. 2018. Bird-building collision risk: An assessment of the collision risk of birds with buildings by phylogeny and behavior using two citizen-science datasets. PLoS ONE 13: e0201558. https://doi.org/10.1371/journal.pone.0201558.

Nielsen, J. 2006. Windows: a clear danger to birds. Morning Edition, National Public Radio, 3 January. Available online at http://www.npr.org/templates/story/story.php?storyId=5076012.

Nisbet, I. C. T. 1970. Autumn migration of the Blackpoll Warbler: evidence for long flight provided by regional survey. Bird Banding 41:207-240.

North American Bird Conservation Initiative. U. S. Committee. 20016. The state of the birds, United States of America. U.S. Department of Interior, Washington, D.C.

Nuttall, T. 1832. A manual of the ornithology of the United States and Canada. Hilliard and Brown, Cambridge, Massachusetts, USA.

Nyffeler, M., C. H. Sekercioglu, and C. J. Whelan. 2018. Insectivorous birds consume an estimated 400-500 million tons of prey annually. The Science of Nature 105:47. https://doi.org/10.1007/s00114-018-1571-z.

Ocampo-Penuela, N., L. Penuela-Recio, and A. Ocampo-Duran. 2015. Decals prevent bird-window collisions at residences: a successful case study from Colombia. Ornitologia Colombiana 15:84-91.

Ocampo-Penuela, N., R. S. Winton, C. J. Wu, E. Zambello, T. W. Wittig, and N. L. Cagle. 2016. Patterns of bird-window collisions inform mitigation on a university campus. PeerJ 4:e1652, DOI:10.7717/peerj.1652.

O'Connell, T. J. 2001. Avian window strike mortality at a suburban office park. The Raven 72(2):141-149.

Odeen, A. and O. Hastad. 2013. The phylogenetic distribution of ultra violet vision in birds. BMC Evolutionary Biology 13:36. http://www.biomedcentral.com/1471-2148/13/36.

Oliveira Hagen, E., O. Hagen, J. D. Ibanez-Alamo, O. L. Petchey, and K. L. Evans. 2017. Impacts of urban areas and their characteristics on avian functional diversity. Frontiers in Ecology and Evolution 5:84.

Ortega-Alvarez, R., and I. MacGregor-Fors. 2009. Living in the big city: effects of urban land-use on bird community structure, diversity, and composition. Landscape Urban Planning 90:189-195.

Oviedo, S., R. M. Menacho-Odio. 2015. Actitud en la preferencia de metodos para evitar el choque de aves contra puertas y ventanas de vidrio en Costa Rica. Zeledonia 19:22-31.

Pain, D. J., A. A. Cunningham, P. F. Donald, J. W. Duckworth, D. C. Houston, T. Katzner, J. Parry-Jones, C. Poole, V. Prakash, P. Round, and R. Timmins. 2003. Causes and effects of temporospatial declines of Gyps Vultures in Asia. Conservation Biology 17:661-671. doi:10.46/j.1523-1739.2003.0174.x.

Palmer, L. 1976. Through a glass darkly, early warning system for birds. Pacific Search 10:10.

Pariafsai, F. 2016. A review of design considerations in glass buildings. Frontiers of Architectural Research 5:171-193. DOI:10.1016/j.foar.2016.01.006.

Parkins, K. L., S. B. Elbin, and E. Barnes. 2015. Light, glass, and bird–building collisions in an urban park. Northeastern Naturalist 22:84-94 DOI 10.1656/045.022.0113.

Pelley, J. 2014. Campus windows save birds, energy. Frontiers in Ecology and the Environment 12:372-375.

Pennington, d. N., J. Hansel, and R. B. Blair. 2008. The conservation value of urban riparian areas for landbirds during spring migration: land cover, scale, and vegetation effects. Biological Conservation 141:1235-1248.

Peron, G. 2013. Compensation and additivity of anthropogenic mortality: life-history effects and review of methods. Journal of Animal Ecology 82:408-417.

Peters, J. 2018. Site of Super Bowl LII is death trap for birds. USA Today. https://www.usatoday.com/story/sports/columnist/josh-peter/2018/01/30/site-super-bowl-2018-death-trap-birds-eagles-patriots-us-bank-stadium/1079934001/.

Peterson, R. T. 1963. The birds. Time, Inc., New York, USA.

Pfennigwerth, S. 2008. Minimising the swift parrot collision threat: guidelines and recommendations for parrot-safe building design.World Wildlife Fund Australia.

Phillips, C. J. 1941. Glass: the miracle maker. Pitman Publishing Corporation, New York, USA.

Piatt, J. F., C. J. Lensink, W. Butler, M. Kendziorek, and D. R. Nysewander. 1990. Immediate impact of the 'Exxon Valdez' oil spill on marine birds. Auk 107:387-397.

Pierce, C. J. 1941. Tall television tower and bird migration. South Dakota Bird Notes 21:4-5.

Pilkington and the flat glass industry. 2010. Pilkington Report. http://www.pilkington.com/pilkington-information/downloads/pilkington+and+flat+glass+industry+2010.htm.

Podulka, S., R. W. Rohrbaugh, Jr., R. Bonney [eds.]. 2004. Handbook of bird biology, 2nd Edition. Cornell Lab of Ornithology, Ithaca, New York, USA.

Ponce, C. J., C. Alonso, G. Argandona, A. Garcia Fernandez, and M. Carrasco. 2010. Carcass removal by scavengers and search accuracy affect bird mortality estimates at power lines. Animal Conservation 13: 603-612.

Poot, H., B. J. Ens, H. de Vries, M. A. H. Donners, M. R. Wernand, and J. M. Marquenie. 2008. Green light for nocturnally migrating birds. Ecology and Society 13:47, online at http://www.ecologyandsociety.org/vol13/iss2/art47/.

Posser, P., C. Nattrass, and C. Prosser. 2008. Rate of removal of bird carcasses in arable farmland by predators and scavengers. Ecotoxicology and Environmental Safety 71:601-608.

Post, S. 1976. Bird migration and window kills. Earlham College, Richmond, Indiana [Unpublished Manuscript].

Primm, S. L., C. N. Jenkins, R. Abell, T. M. Brooks, J. L. Gittleman, L. N. Joppa, P. H. Raven, C. M. Roberts, and J. O. Sexton. 2014. The biodiversity of species and their rates of extinction, distribution, and protection. Science 344. Doi:10.1126/science.1246752.

Pyle, P., and S. Howell. 1997. Identification guide to North American birds: A compendium of information on identifying, ageing, and sexing "near-passerines" and passerines in the hand. Slate Creek Press, Bolinas, California.

Qaiser, F. 2019. Short, glassy buildings are a bird's worst nightmare. Massive Science. https://massivesci.com/articles/bird-window-collisions/.

Raible, Rolf. 1968. Vogelverluste an Glasflachen und methoden zu ihrer Verhutung. Angewandte Ornithologie 3:75-79.

Ramalho, C. E., and R. J. Hobbs. 2011. Time for a change: dynamic urban ecology. Trends in Ecology and Evolution 27:179-188.

Ranford, R. B. and J. E. Mason. 1969. Nocturnal migrant mortality at the Toronto-Dominion Centre. Ontario Field Biologist 23:26-29.

Rawlings, C. M., and D. J. Horn. 2010. Scavenging rates highest at windowed compared to windowless sites at Millikin University in Decatur, Illinois. Illinois State Academy of Sciences 103:153-158.

Rebke, M., V. Dierschke, C. N. Weiner, R. Aumuller, K. Hill, and R. Hill. 2019. Attraction of nocturnally migrating birds to artificial light: The influence of colour, intensity and blinking mode under different cloud cover conditions. Biological Conservation 233:220-227. doi:10.1016/jbiocon.2019.02029.

Rebolo-Ifran, N., A. di Virgilio, and S. A. Lambertucci. 2019. Drivers of bird-window collisions in southern South America: a two scale assessment applying citizen science. Scientific Reports 9:18148. https://doi.org/10.1038/s41598-019-54351-3.

Remsen, J. V., and D. A. Good. 1996. Misuse of data from mist-net captures to assess relative abundance in bird populations. Auk 113:381-398.

Rich, C., and T. Longcore, Editors. 2006. Ecological consequences of artificial lighting. Island Press, Washington, D.C., USA.

Rich, T. D., C. J. Beardmore, H. Berlanga, P. J. Blancher, M. S. W. Bradstreet, G. S. Butcher, D. W. Demarest, E. H. Dunn, W. C. Hunter, E. E. Inigo-Elias, J. A. Kennedy, A. M. Martell, A. O. Panjabi, D. N. Pashley, K. V. Rosenberg, C. M. Rustay, J. S. Wendt, T. C. Will. 2004. Partners in Flight North America Landbird Conservation Plan. Cornell Lab of Ornithology, Ithaca, New York, USA.

Riding, C. S., and S. R. Loss. 2018. Factors influencing experimental estimation of scavenger removal and observer detection in bird-window collision surveys. Ecological Applications 28:2119-2129.

Richardson, W. J. 1990. Timing of bird migration in relation to weather: updated review, in E. Gwinner, Editor. Bird migration. Berlin Heidelberg: Springer, 78-101.

Riding, C. S., T. J. O'Connell, and S. R. Loss. 2020. Building façade-level correlates of bird-window collisions in a small urban area. Condor Ornithological Applications 122:1-14. Doi:10.1093/condor/duz065.

Robbins, C. S., J. R. Sauer, R. S. Greenberg, and S. Droege. 1989. Population declines in North American birds that migrate to the Neotropics. Proceedings National Academy Science 86:7658-7662.

Robbins, C. S., D. K. Dawson, and B. A. Dowell. 1989. Habitat area requirements of breeding forest birds of the Middle Atlantic States. Wildlife Monograph 103.

Robertson, B., G. Kriska, V. Horvath, and G. Horvath. 2010. Glass buildings as bird feeders: Urban birds exploit insects trapped by polarized light pollution. Acta Zoologica Academiae Scientiarum Hungaricae 56:283-292.

Robinson, A. and S. Wilton. 1973. Owl imprints on window panes. Bird Study 20:143-144.

Robinson, W. L. and E. G. Bolen. 1989. Wildlife ecology and management, 2nd Edition, Macmillian Publishing Company, New York, USA.

Roerig, J. 2013. Shadow boxing by birds – A literature study and new data from southern Africa. Ornithological Observations 4:39-68.

Rogers, S. D. 1978. Reducing bird mortality on a college campus in Colorado. C.F.O. Journal 33:3-8.

Rogers, T. 1958. Palouse-Northern Rocky Mountain region. Audubon Field Notes 12:47.

Ronconi, R. A., K. A. Allard, and P. D. Taylor. 2015. Bird interactions with offshore oil and gas platforms: review of impacts and monitoring techniques. Journal of Environmental Management 147:34-35.

Ros, I. G., P. S. Bhagavatula, LinH-T, and A. A. Biewener. 2017. Rules to fly by: pigeons navigating horizontal obstacles limit steering by selecting gaps most aligned to their flight direction. Interface Focus 7:20160093. http://dx.doi.org/10.1098/rsfs.2016.0093.

Rosenberrg, K. V., A. M. Dokter, P. J. Blancher, J. R. Sauer, A. C. Smith, P. A. Smith, J. C. Stanton, A. Panjabi, L. Helft, M. Parr, and P. P. Marra. 2019. Decline of North American avifauna. Science 366:120-124. Doi:10.1126/science.aaw1313.

Ross, R. C. 1946. People in glass houses should draw their shades. Condor 48: 142.

Rossler, M. 2005. Vermeidung von Vogelanprall an Glasflachen. Weitere Experimente mit 9 Markierungstypen im unbeleuchteten Versuchstunnel. Wiener Umweltanwaltschaft. Biologische Station Hohenau-Ringelsdorf. Online at www.windowcollisions.info.

Rossler, M. 2012. Ornilux Mikado. Prufung im Flugtunnel II der Biologischen Station Hohenau-Ringelsdorf; Wiener Umweltanwaltschaft. Download at Http://www.windowcollisions.info/public/vogelanprall-ornilux-mikado_2012.pdf.

Rossler, M. 2013. VERMINDERUNG VON VOGELANPRALLAN GLASFLACHEN PRUFBERICHTABC BIRD TAPE, TESA ® 4593. Prufung im Flugtunnel II der ABiologischen Station Hohenau-Ringelsdorf nach ONR 191040 und unter Einbeziehung von Spiegelungen bei dunklem Hintergrund (WIN-Versuch). Download at http://wuawien.at/images/stories/publikationen/vogelanprall-bird-tape-tesa.pdf.

Rossler, M., 2015. VOGELANPRALL AN GLASFLACHENPRUFBERICHTBIRDPEN® Prufung nach ONR 191040 und WIN-Versuch im Flugtunnel II der Biologischen Station Hohenau-Ringelsdorf. Download at http://wua-wien.at/images/stories/publikationen/pruefbericht-birdpen-2015.pdf.

Rossler, M., W. Laube, and P. Weihs. 2007. Vermeidung von Vogelanprall an Glasflachen. Experimentelle Untersuchungen zur Wirksamkeit von lasmarkierungen unter naturlichen Lichtbedingungen im Flugtunnel II. Biologische Station Hohenau-Ringelsdorf. Online at www.windowcollisions.info.\.

Rossler, M., and W. Laube. 2008. Vermeidung von Vogelanprall an Glasflachen. Farben, Glasdekorfolie, getontes Plexiglas: 12 weitere Experimente im Flugtunnel II. Biologische Station Hohenau-Ringelsdorf. Online at www.windowcollisions.info.

Rossler, M., W. Laube, and P. Weihs. 2009. Avoiding bird collisions with glass surfaces. Experimental investigations of the efficacy of markings on glass panes under natural light conditions in Flight Tunnel II, Final Report, March 2007). BOKU-Met Report 10, ISSN 1994-4179 (print), ISSN 1994-4187 (online). Online at http://www.boku.ac.at/met/report/BOKU-Met_Report_10_online.pdf.

Rossler, M. E. Nemeth, and A. Bruckner. 2015. Glass pane markings to prevent bird-window collisions: less can be more. Biologia 70: 535-541, DOI:10.1515/biolog-2015-0057.

Rossler, M, and T. Zuna-Kratky. 2004. Vermeidung von Vogelanprall an Glasflachen. Experimentelle Versuche zur Wirksamkeit verschiedener Glas- Markierungen bei Wildvogeln. Bilogische Station Hohenau-Ringelsdorf. Online at www.windowcollisions.info.

Roth, T. C. II, s. L. Lima, W. E. Vetter. 2005. Survival and causes of mortality in wintering Sharp-shinned Hawks and Cooper's Hawks. Wilson Bulletin 117:237-244.

Sabo, A. M., N. D. G. Hagemeyer, A. S. Lahey, and E. L. Walters. 2016. Local avian density influences risk of mortality from window strikes. PeerJ 4:e2170; DOI 10.7717/peerj.2170.

Sainsbury, A. W., P. M. Bennett, and J. K. Kirkwood. 1995. The welfare of free-living wild animals in Europe: Harm caused by human activities. Animal Welfare 4:183-206.

San Francisco Planning Department. 2019. Standards for bird-safe buildings. https://sfplanning.org/standards-bird-safe-buildings#about.

Santos, L. P. S., V. F. de Abreu, and M. F. de Vasconcelos. 2017. Bird mortality due to collisions in glass panes on an Important Bird Area of southeastern Brazil. Revista Brasileina de Ornitologia 25:90-101.

Sauer, J. R., J. E. Hines, J. E. Fallon, K. L. Pardieck, D. J. Ziolkowski, Jr., and W. A. Link. 2011. The North American Breeding Bird Survey, results and analysis 1966-2009. United States Geological Survey, Laurel, Maryland, USA, Online at http://www.mbr-pwrc.usgs.gov/bbs/.

Schaub, M. M. Kery, P. Korner, and F. Korner-Nievergelt. 2011. A critique of 'Collisions mortality has no discernible effect on population trends of North American Birds'. PLoS One 6, e24708.

Schifferli, A. 1956. Sichtbarmachen gefahrlicher Fensterflachen fur Vogel. Ornithologische Beobachter 53:108.

Schiffner, I., D. V. Hong, P. S. Bhagavatula, and M. V. Srinivasan. 2014. Minding the gap: in-flight body awareness in birds. Frontiers in Zoology 11:64, DOI:10.1186/s12983-014-0064-y.

Schiffner, I., and M. V. Srinivasan. 2015. Direct evidence for vision-based control of flight speed in Budgerigars. Scientific Reports 2015; 5:10992. doi:10.1038/srep10992.

Schleidt, W., M. D. Shalter, and H. Moura-Neto. 2011. The hawk/goose story: the classical ethological experiments of Lorenz and Tinbergen, revisited. Journal of Comparative Psychology 152:121-133, http://dx.doi.org/10.1037/a0022068.

Schmid, H., w. Doppler, D. Heynen, and M. Rossler. 2012. Vogelfreundliches Bauen mit Glas und Licht. 2., uberarbeitete Auflage. Schweizerische Vogelwarte Sempach. Download at http://www.vogelglas.info/public/voegel_glas_licht_2012.pdf.

Schmitz, J-P. 1969. Vogelverluste an Glasflachen des Athenaums in Luxemburg. Regulus 9:423-427.

Schneider, R. M., C. M. Barton. K. W. Zirkle, C. F. Greene, K. B. Newman. 2018. Year-round monitoring reveals prevalence of fatal bird-window collisions at the Virginia Tech Corporate Research Center. PeerJ 6:e4562. doi:10.7717/peerj4562.

Schotzko, J. 1962. Interrupted migration. Flicker 34:61.

Schramm, M. J. Fiala, T. Noe, P. Sweet, A. Prince, and C. Gordon. 2007. Calls, captures, and collisions: triangulating three census methods to better understand nightly passage of songbird migrants through the Chicago region during May. Meadowlark 16:122-129.

Schultz, Z. M. 1958. North Pacific Coast region. Audubon Field Notes 12:54.

Sealy, S. G. 1985. Analysis of a sample of Tennessee Warblers window-killed during spring migration in Manitoba. North American Bird Bander 10:121-124.

Seewagen, C. L. 2011. A review of experimental methods used to test the effectiveness of bird-deterring glass. American Bird Conservancy. [Unpublished Report].

Seewagen, C. L. and C. Sheppard. 2019. Bird collisions with windows: an annotated bibliography. American Birding Conservancy, Washington, D.C. USA [Unpublished Report].

Seewagen, C. L., C. D. Sheppard, E. J. Slayton, and C. G. Guglielmo. 2011. Plasma metabolites and mass changes of migratory landbirds indicate adequate stopover refueling in a heavily urbanized landscape. Condor 113:284-297.

Seewagen, C. L., and E. J. Slayton. 2008. Mass changes of migratory landbirds during stopovers in a New York City park. Wilson Journal of Ornithology 120:296-303.

Seewagen, C. L., E. J. Slayton, and C. G. Guglielmo. 2010. Passerine migrant stopover duration and spatial behavior at an urban stopover site. Acta Oecologica 36:484-492.

Sekercioglu, C. H. 2006. Increasing awareness of avian ecologifcal function. Trends in Ecology and Evolution 21:8. doi:10.1016/j.tree.2006.05.007.

Sekercioglu, C. H., G. C. Daily, and P. R. Ehrlich. 2004. Ecosystem consequences of bird declines. Proceedings of National Academy of Sciences 101:18042-18049. doi:10.1073/pnas.0408049101.

Sekercioglu, C. H., D. G. Wenny, and C. J. Whelan. 2016. Why birds matter. University of Chicago Press, Chicago, Illinois, USA.

Seto, K. C., R. Sanchez-Rodriguez, and M. Fragkias. 2010. The new geography of contemporary urbanization and the environment. Annual Review Environmental Resources 35:167-194.

Seto, K. C., B. Guneralp, and L. R. Hutyra. 2012. Global forecasts of urban expansion to 2030 and direct impacts on biodiversity and carbon pools. Proceedings of the National Academy of Sciencs USA 109:16083-16088.

Sheppard, C. 2019. Evaluating the relative effectiveness of patterns on glass as deterrents of bird collisions with glass. Global Ecology and Conservation 20 e00795. https://doi.org/10.1016/j.gecco.2019.e00759.

Sheppard, C., and G. Phillips. 2015. Bird-friendly building design, 2nd Edition. American Bird Conservancy, The Plains, Virginia, USA, Online at http://collisions.abcbirds.org.

Sierro, A., and H. Schmid. 2001. Impact des vitres transparentes antibruit sur les oiseaux: une saison d'experience a Brig VS. In: Actes du 39e colloque interregional d'ornithologie, vol. Suppl 5, Yverdon-les-Bains (Suisse), Nos Oiseaux. Pp. 139-143.

Sinner II, P. F. 1972. Window kill research project. Valley City State College, Valley City, North Dakota, USA [Unpublished Manuscript].

Skutch, A. F. 1977. A bird watcher's adventures in tropical America. University of Texas Press, Austin, Texas, USA.

Sloan, A. 2007. Migratory bird mortality at the World Trade Center and World Financial Center, 1997-2001: a deadly mix of lights and glass. Transactions of the Linnaean Society of New York 10:183-204, Download at http://linnaeannewyork.org/Transactions%20X.pdf.

Smallwood, K. S. 2007. Estimating wind turbine-caused bird mortality. Journal of Wildlife Management 71:2781-2791.

Smallwood, K. S. and C. Thelander. 2008. Bird mortality in the Altamont Pass Wind Resource Area, California. Journal of Wildlife Management 72:215-223.

Smith, K. A., G. D. Campbell, D. L. Pearl, C. M. Jardine, F. Salgado-Bierman, and N. M. Nemeth. 2018. A retrospective summary of raptor mortality in Ontario, Canada (1991–2014), including the effects of West Nile Virus. Journal of Wildlife Diseases 54:1-11. DOI:10.7589/2017-07-157.

Snep, R. P. H., J. L. Kooijmans, R. G. M. Kwak, R. P. B. Foppen, H. Parsons, M. Awasthy, H. L. K. Sierdsema, J. M. Marzluff, E. Fernandez-Juricic, J. de Laet, and Y. M. van Heezik. 2016. Urban bird conservation: presenting stakeholder-specific arguments for the development of bird-friendly cities. Urban Ecosystems 19:1534. DOI:10.1007/s11252-015-0442-z.

Snyder, L. L. 1946. "Tunnel fliers" and window fatalities. Condor 48:278.

Somerlot, D. E. 2003. Survey of songbird mortality due to window collisions on the Murray State University campus. Journal of Service Learning in Conservation Biology 1:1-19.

Sovacool, B. 2009. Contextualizing avian mortality: A preliminary appraisal of bird and bat fatalities from wind, fossil-fuel, and nuclear electricity. Energy Policy 37:2241-2248. DOI.10.1016/j.enpol.2009.02.011.

Stedman, S. J., and B. H. Stedman. 1986. Preventing window strikes by birds. Migrant 57:18.

Steffen, W., J. Crutzen, and J. R. McNeill. 2007. The Anthropocene: Are humans now overwhelming the great forces of Nature? Ambio 36:614-621. doi:10.1579/0044-7447(2007)36[614.TAAHNO]2.0CO.

Stevens, B. S., K. P. Reese, and J. W. Connelly. 2011. Survival and detectability bias of avian fence collision surveys in sagebrush steppe. Journal of Wildlife Management 75:437-449.

Stevens, M. 2011. Avian vision and egg colouration: concepts and measurement. Avian Biology Research 4(4):168-184."

Stevens, M. C. A. Parraga, I. C. Cuthill, J. C. Partridge, and T. S. Troscianko. 2007. Using digital photography to study bird coloration. Biological Journal of the Linnean Society 90:211-237.

Stoddard, M. C., H. N. Eyster, B. G. Hogan, D. H. Morris, E. R. Soucy, and D. W. Inouye. 2020. Wild hummingbirds discriminate nonspectral colors. Proceedings of the National Academy of Sciences 117:15112-15122. https://doi.org/10.1073/pnas.1919377117.

Stracey, C. M., and S. K. Robinson. 2012. Is an urban-positive species, the Northern Mockingbird, more productive in urban landscapes? Journal of Avian Biology 43:50-60.

Stratford, J. A., W. D. Robinson. 2005. Distribution of Neotropical migratory birds across an urbanizing landscape. Urban Ecosystems 8:59-77.

Stutchbury, B. 2007. Silence of the songbirds. Walker and Company, New York, USA.

Swaddle, J. P., C. D. Francis, J. R. Barber, C. B. Cooper, C. C. M. Kyba, D. M. Dominoni, G. Shannon, E. Aschelhoug, S. E. Goodwin, A. Y. Kawahara, D. Luther, K. Spoelstra, M. Voss, and T. Longcore. 2015. A framework to assess evolutionary responses to anthropogenic light and sound. Trends in Ecology and Evolution 30:550-560.

Swaddle, J. P. and N. M. Ingrassia. 2017. Using a sound field to reduce the risks of bird-strike: An experimental approach. In Symposium: Indirect effects of global change: from physiological and behavioral mechanisms to ecological consequences (SICB wide. Integrative and Comparative Biology, pp. 1-9. Doi:10.1093/icb/icx026.

Swaddle, J. P., L. C. Emerson, R. G. Thady, T. J. Boycott. 2020. Ultraviolet reflective film applied to windows reduces the likelihood of collisions for two species of songbirds. PeeJ 8:e9926 DOI 10.7717/peerj.9926.

Switala Elmhurst, K., and K. Grady. 2017. Fauna protection in a sustainable university campus: bird-window collision mitigation strategies at Temple University, in W. Leal Filho, L. Brandli, P. Castro, and J. Newman, Editors. Springer International, 69-82. Online at https://sustainability.temple.edu/birds.

Tallarico, R. B. and W. M. Farrell. 1964. Studies of visual depth perception: an effect of early experience on chicks on a visual cliff. Journal of Comparative Physiological Psychology 57:94-96.

Talpin, A. 1991. A little used source of data on migrant birds. Corella 15:24-26.

Tan, D. 2018. Birdstrike deterrent through community-based application of oil-paint markers on glass. Biodiversity Research Center, Beaty Biodiversity Museum, University of British Columbia.

Taylor, H. M. 2003. Preventing avian mortality due to window collisions. Department of Biology, University of California, Santa Cruz. [Unpublished Manuscript].

Taylor, W. K., and M. A. Kershner. 1986. Migrant birds killed at the vehicle assembly building (VAB), John F. Kennedy Space Center. Journal of Field Ornithology 57:142-154.

Temple, S. A. and J. A. Wiens. 1989. Bird populations and environmental changes: can birds be bio-indicators? American Birds 43:260-270.

Terres, J. K. 1980. The Audubon Society encyclopedia of North American birds. Alfred A. Knopf, New York, New York, USA.

Tiemeier, O. W. 1941. Repaired bone injuries in birds. Auk 58:350-359.

Townsend, C. W. 1931, Tragedies among Yellow-billed Cuckoos. Auk 48:602.

Troscianko, J., and M. Stevens. 2012. Image calibration and analysis toolbox – a free software suite for objectively measuring reflectance, colour and pattern. Methods in Ecology and Evolution. https://doi.org/10.1111/2041-210X.12439.

Trybus, T. 2003. Wirksamkeit von Greifvogelsilhouetten zur Verhinderung von Kleinvogelanprall an Glasfronten. Die These des Masters, der Universitat Wien, Austria.

Underwood, A. J. 1992. Beyond BACI: The detection of environmental impacts on populations in the real, but variable, world. Journal of Experimental Marine Biology and Ecology 161:145-178. DOI:10.1016/0022-0981(92)90094-Q.

United States Congress. 2019. H.R.919 – Bird-Safe Buildings Act of 2019. https://www.congress.gov/bill/116-congress/house-bill/919/text.

United States Department of the Interior, U.S. Fish and Wildlife Service, and U. S. Department of Commerce, U.S. Census Bureau. 2011. National survey of fishing, hunting, and wildlife-associated recreation. Washington, D.C., USA.

United States Fish and Wildlife Service. 2008. Birds of conservation concern 2008. U.S. Department of Interior, Fish and Wildlife Service, Division of Migratory Bird Management, Arlington, Virginia, USA, https://www.fws.gov/migratorybirds/pdf/grants/BirdsofConservationConcern2008.pdf.

United States Fish and Wildlife Service. 2016. Reducing bird collisions with buildings and building glass best practices. Division of Migratory Bird Management, Falls Church, Virginia, USA.

US Green Building Council. 2019a. LEED v4 for building design and conservation. Washington, D.C.: US Green Building Council.

U.S. Green Building Council. 2019b. Pilot-credits SSpc55: bird collision deterrence. Washington, D.C.: US Green Building Council.

Valum, B. 1968. Fugledod mot glassvegger. Sterna 8:15-20.

Van Doren, B. M., K. G. Horton, A. M. Dokter, H. Klinck, S. B. Elbin, and A. Farnsworth. 2017. High intensity urban light installation dramatically alters nocturnal bird migration. Proceedings of the National Academy of Sciences. doi:10.1073/pnas.1708574114.

Veltri, C. J. and D. Klem, Jr. 2005. Comparison of fatal bird injuries from collisions with towers and windows. Journal of Field Ornithology 76:127-133.

Verheijen, F. J. 1958. The mechanism of trapping effect of artificial light sources upon animals. Archives Neerlandaises de Zoologie 13:1-107.

Verheijen, F. J. 1980. The moon: a neglected factor in studies on collisions of nocturnal migrant birds with tall lighted structures and with aircraft. Die Vogelwarte 30:305-320.

Verheijen, F. J. 1981a. Bird kills at lighted man-made structures: not on nights close to a full moon. American Birds 35:251-254.

Verheijen, F. J. 1981b. Bird kills at tall lighted structures in the USA in the period 1935-1973 and kills at a Dutch lighthouse in the period 1924-1928 show similar lunar periodicity. Ardea 69:199-203.

Verheijen, F. J. 1985. Photopollution: artificial light optic spatial control systems fail to cope with incidents, causations, remedies. Experimental Biology 44:1-18.

Vitala, J., E. Korpimaki, P. Palokangas, and M. Koivula. 1995. Attraction of kestrels to vole scent marks in ultraviolet light. Nature 373:425-427.

Vo, H. D., I. Schiffner, and M. V. Srinivasan. 2016. Anticipatory manoeuvres in bird flight. Nature: Scientific Reports 6:27591. doi:10.1038/srep27591.

Vorobyev, M., D. Osorio, A. T. D. Bennett, N. J. Marshall, and I. C. Cuthill. 1998. Tetrachromacy, oil droplets and bird plumage colours. Journal of Comparative Physiology A 183:621-633.

Vucetich, J. A., and M. P. Nelson. 2007. What are 60 warblers worth? Killing in the name of conservation. Oikos 116:1267-1278.

Walk, R. D. and E. a. Gibson. 1961. A comparative and analytical study of visual depth perception. Psychological Monograph 75(15 Whole No. 519).

Walkinshaw, L. R. 1976. A Kirtland's Warbler life history. American Birds 30:773.

Wallace, G. J. and H. d. Mahan. 1975. An introduction to ornithology, 3rd Edition, MacMillan Publishing Company, Inc., New York, USA.

Walls, G. L. 1942. The vertebrate eye and its adaptive radiation. Cranbrook Institute of Science, Bloomfield Hills, Michigan, USA.

Ward, M. R., D. E. Stallknecht, J. Willis, M. J. Conroy, and W. R. Davidson. 2006. Wild bird mortality and West Nile virus surveillance: biases associated with detection, reporting, and carcass persistence. Journal of Wildlife Diseases 42:92-106.

Watson, M. J., D. R. Wilson, and D. J. Mennill. 2016. Anthropogenic light is associated with increased vocal activity by nocturnally migrating birds. Condor 118:338-344. doi:10.1650/CONDOR-15-136.1.

Webster, F. J. 1965. Spring migration: south Texas region. Audubon Field Notes 19:490-497.

Wedeles, C. H. R. 2010. Avian incidental take due to buildings in Canada. Report prepared by ArborVitae Environmental Services Ltd. For Environment Canada [Unpublished Manuscript].

Weir, R. D. 1976. Annotated bibliography of bird kills at man-made obstacles: a review of the state of art and solutions. Department of Fisheries and the Environment. Environmental management Service, Canadian Wildlife Service Ontario Region.

Weise, C. M. 1979. Sex identification in Black-capped Chickadees. Field Station Bulletin 12:16-19.

Weisman, A. 2007. The world without us. Thomas Dunne Books, St. Martin's Press, New York, USA.

Welty, J. C. 1975. The life of birds, 2nd Edition, W. B. Saunders Company, Philadelphia, Pennsylvania, USA.

Whelan, P. 1976. The bird killers. Ontario naturalist 16:14-16.

Whelan, C. J., C. H. Sekercioglu, and D. G. Wenny. 2015. Why birds matter: From economic ornithology to ecosystem services. Journal of Ornithology 156:227-238. doi:10.1007/s10336-015-1229-y.

White, R. L., A. E. Sutton, R. Salguero-Gomez, T. C. Bray, H. Campbell, E. Cieraad, N. Geekiyanage, L. Gherardi, A. C. Hughes, and P. S. Jorgensoen. 2015. The next generation of action ecology: Novel approaches towards global ecological research. Ecosphere 6:1-16. DOI:10.1890/ES14-00174.1.

Whittaker, K. A. and J. M. Marzluff. 2009. Species-specific and relative habitat use in an urban landscape during the postfledging period. Auk 126:288-299.

Wiese, F. K., W. A. Montevecchi, G. K. Davoren, F. Huettmann, A. W. Diamond, and J. Linke. 2001. Seabirds at risk around offshore oil platforms in the North-west Atlantic. Marine Pollution Bulletin 42:1285-1290. Download at http://play.psych.mun.ca/~mont/pubs.html.

Willett, G. 1945. Does the Russet-backed Thrush have defective eyesight? Condor 47:216.

WindowAlert.2012. WindowAlert decals. Online at http://wwwwindowalert.com/.

Winger, B. M., B. C. Weeks, A. Farnsworth, A. W. Jones, M. Hennen, and D. E. Willard. 2019. Nocturnal flight-calling behavior predicts vulnerability to artificial light in migratory birds. Proceedings Royal Society B 286:20190364. http://dx.doi.org/10.1098/rspb.2019.0364.

Winker, K., D. W. Warner, and A. R. Weisbrod. 1991. Unprecedented stopover site fidelity in a Tennessee Warbler. Wilson Bulletin 103:511-512.

Winton, R. S., N. Ocampo-Penuela, and N. Cagle. 2018. Geo-referencing bird-window collisions for targeted mitigation. PeerJ. DOI 10.7717/peerj.4215.

Withgott, J. 2000. Taking a bird's-eye view ... in the UV. Bioscience 50:854-859.

Wittig, T. W., N. L. Cagle, N. Ocampo-Penuela, R. Scott Winton, E. Zambello, and Z. Lichtneger. 2017. Species traits and local abundance affect bird-window frequency. Avian Conservation and Ecology 12:17. https://doi.org/10.5751/ACE-01014-120117.

Witting, T. 2016. New perspectives on bird-window collision: the effects of species traits and local abundance on collision susceptibility. Master's Thesis, Duke University, Durham, North Carolina, USA. Online at http://dukespace.lib.duke.edu/dspace/handle/10161/11898.

Witzler, S., M. Van Hoose, A. Hyder, S. Heywood, L. Geroge, and J. Clark. 1976. A five week study on window kills during spring migration at Earlham College, Richmond, Indiana. [Unpublished Manuscript].

Wood, J. S. 2014. Birds, buildings and LEED mitigation design at the University of Calgary campus. Thesis. University of Calgary. Calgary, Alberta, Canada.

Woodrey, M. S., and C. R. Chandler. 1997. Age-related timing of migration: Geographic and interspecific patterns. Wilson Bulletin 109:52-67.

Wulf, A. 2015. The invention of nature Alexander Von Humboldt's new world. Vintage Books, A Division of Penguin Random House LLC, New York.

Yakutchik, M. 2003. Fatal reflections. Philadephia Inquirer Magazine, 11 May:12-17.

Yanagawa, H., and T. Shibuya. 1998. Causes of wild bird mortality in eastern Hokkaido. III. Bird-window collisions. Research Bulletin of Obihio University 20:253-258.

Young, D. P., Jr., W. P. Erickson, M. D. Strickland, R. E. Good, and K. J. Sernka. 2003. Comparison of avian responses to UV-light-reflective paint on wind turbines. Subcontract Report July 1999–2000. National Renewable Energy Laboratory, Golden, Colorado, USA.

Youth, H. 2003. Winged messengers: The decline of birds. Worldwatch Institute, Worldwatch Paper 165.

Zbyryt, A., A. Suchowolec, and R. Siuchno. 2012. Species composition of birds colliding with noise barriers in Bialystok (North-Eastern Poland). International Study of Sparrows 36:88-94.

Zeigler, H. P. and H-J Bischof, Editors. 1993. Vision, brain, and behavior in birds. A Bradford Book, The MIT Press, Cambridge, Massachusetts, USA.

Zink, R. M., and J. Eckles. 2010. Twin Cities bird-building collisions: a status update on "Project Birdsafe." Loon 82:34-37.

Zysk-Gorczynska, E., P. Skorka, and M. Zmihorski. 2019. Grafiti saves birds: A year-round pattern of bird collisions with glass bus shelters. Landscape and Urban Planning 193. http://doi.org/10.1016/j.landurbplan.2019.103680.

Acknowledgments

I AM INDEBTED AND GRATEFUL to the following principal mentors who directly or indirectly fostered and improved by understanding of birds and scientific research: Terrence R. Anthoney, Richard C. Banks, Donald L. Beggs, Ronald A. Brandon, Paul A. Buckley, William G. Dyer, Paul R. Earl, William G. George, Herman Haas, Willard D. Klimstra, Eugene A. LeFebvre, Charleton J. Phillips, Charles B. Reif, Efrem Rosen, Howard A. Swain, Jr., and George H. Waring.

I AM AS INDEBTED AND grateful to my research students and others who directly or indirectly, modestly or extensively, meaningfully contributed to my bird-window studies and enriched my life. They are: Susan Ackermann, Ian Adler, Austin D. Akner, Andreas Asch, Amy J. (Miller) Ball, Phillip J. Baney, IV, Karen (Cotter) Bennett, Kara E. Blum, William E. Bowman, Chris R. Brancato, Clait E. Braun, Jon E. Brndjar, Brandon Brogle, Barbara Buchinski, Joseph F. Catalano, John E. Chojnowski, Craig A. Clifford, Randy Comeleo, Louis S. Crivelli, II, Andrew Curley, Glen A. Davis, Viviane C. Dehmel, William John Deibler, II, Robert J. Driver, Kate Ekanem, Margaret Elder, Sarah M. Elrod, Mary Kate Erdman, Lynne L. Fallon, Laura Farpour, Alex G. Fetterman, Susan A. Finn, Steve Fluck, Michelle Frankel, James W. Freeman, Vincent P. Fugazzotto, Paul J. Galgano, Marion E. Glick, Miguel Angel Gomez, John D. Goodrich, Ryan T. Gross, Vanden L. Grube, Paula J. Halupa, Karin Haney, Adam Heide, Rebecca Hernandez, Heidi Hessenthaler, Brian A. Hillegass, Dayna C. Hovern, David C. Keck, Karen L. Keck, Michael Keogh, Scott L. Kleinberg, Andrew J. Kleiner, Erin V. Kreider, Elianna G. Kronisch, Joy Kuchler, Debra Kupcha, Stephen B. Kupferberg, Frank L. Kuzmin, Rebecca L. Lee, Bradley Leitgeb, French A. Lewis, III, Katarina M. Liberatore, Emma K. Loh, Eric Luthi, Vicki L. Marks, Karl L. Marty, Michael R. Marvin, Christian Mathers, Natalie May, Timothy R. McCreesch, Timothy R. McKnight, Amy J. Miller, Luke C. Miller, Luke A. Munyan, Mathew K. Myers, Richard Nash, Joseph H. Nave, Jr., Richard Nelson, Elizabeth E. Niciu, John Ochsenreither, Rose Marie Menacho Odio, Melanie A. Parker, Diane A. Peters, Joshua Phillips, John R. Pisciotta, Corry T. Platt, Dennis T. Punyodyana, Joseph Purcell, Scott W. Pyne, Sara A. Reibschied, Stacy D. Romascavage,

David L. Rosolia, Michelle G. Rossi, Gena L. Rudikoff, Graceanne S. Ruggiero, Thomas J. Sabo, Geofrey H. Saunders, Nicholas Savant, Spencer Schock, John F. Schumacher, Chad Schwartz, Gerald D. Scott, Claudia Seyfert, Louis E. Spikol, William L. Sprague, Paul T. Stathis, Patrick G. Stockler, Jason R. Stout, Scott A. Tafuri, Bruce Terry, Eric Van Tol, Robert J. Torphy, Heidi Trudell, Nicholas M. Varvarelis, Carl J. Veltri, Joseph A. Villa, Kimberly S. Ware, Brett Weber, James S. Whitaker, Richard S. Wilson, Dorothea Yialamas, Thomas S. Ziering,

I THANK THE FOLLOWING SKILLED craftsmen and craftswomen, building industry professionals (architects, developers, engineers, glass manufacturers, landscape designers and architects), conservation and ornithology scientists and citizen-scientists, and writers who directly or indirectly, modestly or extensively, also meaningfully helped me in my research and shared their knowledge about bird-window interactions. They are: David Aborn, Jeff Acopian, Sarkis Acopian, Beth Adams, Claude Adams, Lowell W. Adams, Paul S. Adams, Natia Albramia, Charles Alexander, Stephanie Allard, Debra Allen, Robert (Bob) Alsip, Leigh Altadonna, Marissa C. G. Altmann, Stephen Ambrose, Bruce D. Anderson, George W. Archibald, Hans-Joachim Arnold, Bob Arsenault, Mathew E. Assad, Erik C. Atkinson, Maggie Atmish, Michael L. Avery, E. A. (Andy) Axtell, III, Nadia A. Ayoub, Jerry Bahls, Paul J. Baicich, Susan Ban, John Barry, Joe Bartell, Ellen Bartyzal, Denise Bauer, Sandy Bauers, Christopher Baxter, Jeff Baxter, Robert Beason, Brenda Beatty, Sarah Beazley, Paula Beck, Bonnie Belknap, Chris Bennett, Cynthia Berger, Bonnie Bernstein, Laura Bies, Keith L. Bildstein, George Black, Jr., Charles R. Blem, Leann Blem, Giovanni Boano, Madeline Bodin, Henry Bokuniewicz, Walter Bonner, Ben Borden, Sandy Bourque, David Brandes, Laurent Brasier, Clait E. Braun, Daniel W. Brauning, Greg Brecht, Raymond Brereton, James J. Brett, Allan Britnell, Michael Broekhuis, Howard P. Brokaw, Bernard Brown, Kyle G. Brown, Mary Bomberger Brown, Theresa Buckhus, Michael Bruckner, Francine Buckley, Adam Burghardt, Scott Burnet, Brooks M. Burr, Edward (Jed) H. Burtt, Greg Butcher, Michael Butler, Michael W. Butler, Rhett Butler, Paul Cabe, Scott R. Campbell, Kelly Cannon, Peter Cannon, John Robert Carley, J. Frederick Charbonneau, Jimmy Charter, Kay Charter, Jameson F. Chase, Kit Chubb, J. Alan Clark, Ben Coleman, Mary Coolidge, Sean Cooney, Anthony Corazza, Kari (P'Brien) Corazza, Ray Corwin, Karen Cotton, David Crawford, Rebekah Creshkoff, Kevin F. Crilley,

Acknowledgments

Daniel A. Cristol, James Cubie, Robert L. Curry, Ken Cwalina, Richard M. Daley, Molly Daly, Danik Dancause, Donnie Dann, Anne Davis, Ed Davis, Glenn De Baeremaeker, Krista De Groot, Jeff de Waal, Robert DeCandido, Nicole Delacretaz, Jeff Demko, Lyman C. Dennis, Marc Deschamps, Alice Deutsch, Scott Diehl, Jack Dillon, Marco Dinetti, Randi B. Doeker, Carla J. Dove, Chris Draper, James E. Ducey, Jim Duffy, Erica H. Dunn, Peter Dunne, Alice During, Kent Dyer, Susan B. Eblin, Joanna Eckles, Kevin Eddings, David Edger, Anne Eisenberg, Kate Ekanem, Sarah Elston, Eleanor Emanuel, Mary Erdman, Laura Erickson, Bill Evans, Lesley Evans-Ogden, E. Carr Everbach, Nicoletta Fabiani, David Fagerstrom, Christopher J. Farmer, Irene Fedun, Tom Fegley, Jason Feiertag, Rob Fergus, Mike Ferry, Rachel Fetherston, Wolfgang Fiedler, Stephanie Findlay, Erica Finkelstein, James Karl Fisher, Sarah Flournoy, Lisa W. Foderaro, Tom Fontana, Steve P. Fordyce, Bruce Fowle, Marcia Fowle, Carol Frank, Stephen Frantz, David Fraser, Dyana Z. Furmansky, Edwin C. Galbreath, Rose Gallagher, Jeanne Gang, Kellie Patrick Gates, Alma Gaul, Joelle Gehring, Yigal Gelb, Ben Gelman, Marian George, Shara George, Frank B. Gill, Andrea Gillespie, Edward (Ted) S. Gilman, Matt Giza, Abigail Golden, Greg Goldman, L. A. Sanchez Gonzalez, Laurie J. Goodrich, Jean W. Graber, Richard R. Graber, Duane D. Gray, Michael Green, Tina Green, Russell S. Greenberg, Stephen Greenfield, Anton Greenwald, Jamie Greiss, Vince Grippo, Tyler Groft, Paul Groleau, Doug Gross, Frank Haas, John Hadidian, Julie Hagelin, Stephen B. Hager, Georgia Hariton, Lee Harrison, Michael Harwood, Jerry Hassinger, Robert (Bob) Hastings, Heiko Haupt, Ruth Heil, Nancy Held, Margaret Helfand, Peg Hentz, Paul Hess, Thomas Hicks, Joyce Hinnefeld, William Hodos, Paquita Hoeck, Anouk Hoedeman, Stephen W. Hoffman, Marguerite Y. Holloway, Vicki Holmberg, Jonathan Holtzman, Michael (Mike) Homoya, David J. Horn, Jack Hubley, Robbie Lynn Hunsinger, Simon Hurst, Catherine Ingram, Christian Irmscher, Jerome A. Jackson, Joanne Jackson, Rebecca Jasulericz, Andrew Jenner, Janie Johns, Andy Jones, Christopher (Chris) Joyce, Edward (Ted) Fogg Judt, Ron L. Kagan, Bettina Kain, Kiki Keating, Brandon Keim, Ronald (Ron) Keiper, Rudy Keller, William (Bill) Keller, Alexander Kelly, Paul Kerlinger, Jacob Kichline, Alexander Kienler, Young-Jun Kim, Timothy Kimmel, Alicia F. (Craig) King, Iochem Kirchner, Karen Knee, Albert Koehl, Julia Koch, Casey Kosco, Susan Krajnc, Stephen W. Kress, Anne M. Kuntz, Dan Kunkle, Robert Kuper, Ken Lauer, Marjorie H. Lauer, Anne Laughlin, Virginia (Jenny) Lecko, Vladimir (Walter) Lecko, David S. Lee, Jane J. Lee, Sugil Lee, Waiman Lee, Daniel L.

Leedy, Julie Leibach, Wendy Lenhart, Pat Leonard, Sherona Leung, Scott Levinson, Aidan Lewis, Anne Lewis, David Lewis, Tracy Librick, Manfred Lieser, Janea Little, Travis Longcore, Scott R. Loss, Erika Lovejoy, JoAnn Loviglio, Michael Lundin, Tom Lyons, Ian MacGregor-Fors, Craig Machtans, Andrew L. Mack, Barbara Mahany, Ken Mahmud, Shyamal K. Majumdar, David Malakoff, Barbara Malt, Jeff Manning, Albert M. Manville, II, Carroll Marino, Frank Marino, Peter P. Marra, Helene Marshall, Graham R. Martin, Michael Martin, Tim Martin, Terry L. Master, Guy Maxwell, Steven G. Mayer, E. J. McAdams, James McCommons, D. J. McCutheon, Charlotte McDonald, Matt McFarland, Chris McGrory, Mike McGrory, Dennis McNair, Karen McNair, Neil McSporran, Paul T. Meier, David J. Mellor, Amanda (Mandy) Meltz, Liz Merfeld, Michael Mesure, Alfred Meyerhuber, Susan Milius, Jean Minnich, Michael M. Miorelli, Kees Moeliker, Fabiano Montiani-Ferreira, Charisa Morris, Ruth Morris, Diane Motel, Angelo M. (Mike) Mucci, Anthony (Tony) M. Muir, Jo Jo Muir, Helene Munch, Jeff Naatz, Amy Neale, David Neale, Linda Nemes, Nicholas Newberry, Marie A. North, Daniel Novak, Gary L. Nunn, Pamela G. Nunn, George E. O'Donnell, Wendy Olsson, Alisa Opar, Carl S. Oplinger, Stephen B. Oresman, Dianna Ortiz, Francisco Ortiz, Henri Roger Ouellet, Chris Ouellette, Jonathan Panigas, Kaitlyn L. Parkins, Lynne Parks, Mary Lou Parlato, Phillip C. Parsons, Robert O. Paxton, Mark K. Peck, Valerie Peckham, Brian Pedersen, Constance Pepin, Wayne R. Petersen, Glenn Phillips, Thomas Pinnekamp, Agusto Piratelli, Dan Pisselli, Paloma Plant, Mark Pokras, Matt Polis, Dianne Pollock, Judy Pollock, Anthony (Tony) Port, Ellen Pothering, Rene Potter, Annette Prince, David Rabold, Rolf Raible, Daniel Ramp, William (Will) Rantz, Wilma E. A. Reese, Ulrich Reich, Frank Riccio, Terrell (Terry) D. Rich, Austin Richards, Madeleine (Maddy) Richards, Wendy Richards, Bruce Riddle, Corey Riding, Greg Rienzi, Bruce Robertson, Vicki Rock, Bjoern Rosen, Laurie Rosenberg, Tamara Rosenberg, Judith (Joy) Ross, Martin Rossler, Polly Rothstein, Eitan Rudski, Gloria Rudski, Jeff Rudski, William (Bill) Ruhe, Karen Carlo Ruhren, Ric Rupnik, Jeff Rusinow, Keith Russell, Meghan Sadlowski, Lee Ann Saenger, Tina Saylor, Robert L. Schaeffer, Jr., Jeremy Scheivert, Mark Schendel, Jonathan Schlegel, Wolfgang Schleidt, Jean Yockey Schmoyer, Irvin R. Schmoyer, Rebecca Murray Schneider, Spenser Schock, Manfred Schon, Hal Schramm, Chad L. Seewagen, Amy Serafin, Lis Shapiro, Christine Sheppard, Gayland Shirk, Warren Shirk, Deepak Shivaprasad, Carol Sholl, David Sibley, James C. Simmons, Alan C. Simon, Lee Simpson, Maqbool A. Siraj, Marc

Acknowledgments

Sklar, Simon Slade, James Smith, Kelly Snow, Alexander Sobolev, Vivienne Sokol, Rachel I. (Taras) Spagnola, Helen Spengler, Hari Sreenivasan, Nicholas St. Fleur, Sarah Staffiere, Amy Standen, Michael A. Steele, Grant Stevenson, Jamie Stover, David Stricker, D. H. Strongheart, Meera Subramanian, Denis Sylvain, Howard Szczech, Tina Tam, Paul Tankel, Lou Tanner, Stephanie Tanner, Andrew T. Tax, Lenore P. Tedesco, Karen Terbush, Monica Tomosy, Laura June Topolsky, Maya Totman, William (Bill) Trachtenberg, Michael Traupman, Peter Tung, George Turjanica, John Turner, Bonnie Van Dam, Robert (Bob) Vanderjack, Mark Van Tuinen, Alexander von Mezynski, Robert Vandal, James R. Vaughan, Joanne Evans Vaughan, Victor Veerasamy, Tabitha Viner, Daniel Viola, Peter Virgili, Ingrid Visser, Dominique Waddoup, Anna Wade, Christina Wallitsch, Frank Warendorf, Ann-Meredith Waring, David Warwick, Francis (Frank) D. Watson, David Webber, Lisa Welch-Schon, Jiangang Weng, Urban (Brother)Wentling, Cody Wessner, Steve Wessner, John C. Weston, Ben Whitford, Richard J. Whittington, Bruce C. Wightman, Edward (Ned) H. Williams, Timothy C. Williams, Herb Wilson, Rick Wiltraut, Keith Winsten, Philip Witmer, Sr., Roger C. Wood, Sarah Wren, Jim Wright, Francis Wuillaume, Kurt D. Yalcin, Clara Yambert, Laura Yambert, Paul Yambert, Jim Yukes, Adam Zbyryt, Anthony Zemba, Xia Zhang.

I AM ESPECIALLY THANKFUL TO the hundreds of individuals, far too many to list here, who recorded and provided me with documentation of species striking windows the world over.

I AM GRATEFUL TO THE following architecture, commercial, conservation, education, engineering, legal, manufacturing, and research institutions who directly or indirectly, modestly or extensively, offered speaking opportunities and programs, products and funding that supported my studies to better understand bird-window collisions and how to prevent them. They are: 3M, Acopian Technical Company, Acopian BirdSavers, AGC North America, American Bird Conservancy (ABC), Ambrose Ecological Services, American Museum of Natural History, American Ornithologist's Union (American Ornithological Society), Animal Help Now, Apple Technology Company, Arnold Glas, Artscape Inc., Association of Field Ornithologists, Audubon Chapter of Minneapolis, Audubon Greenwich, Audubon Minnesota, Audubon Pennsylvania, Augustana College, Baird

Ornithological Club, Baltimore Bird Club, Berks County Community – TV, Bird Alert, Birds and Building Forum, Bloomsburg University, Born Free Foundation, British Broadcasting Company, Butler University, Canadian Wildlife Service, Carnegie Museum of Art and Natural History, Chautauqua Institute, Chicago Bird Collision Monitors, Chicago Ornithological Society, Chippewa Nature Center, Civic Theatre of Allentown, CollidEscape/Bierte Inc., City of Allentown, City of Calgary, City of Portland, City of San Francisco, City of Toronto, College of William and Mary, Colby College, Connecticut Audubon Society, Connecticut Ornithological Association, Convenience Group – Feather Friendly, Cornell Lab of Ornithology, Costa Rica Ornithological Society, CP Films (Eastam Chemical), Curry and Kerlinger LLC, Da Vinci Science Center, Delaware Ornithological Club, Delaware Museum of Natural History, Detroit Audubon Society, Detroit Zoological Society, e-Time Energy, East Stroudsburg University, Eastern Bird Banding Association, Ecojustice, Erickson International LLC, Fatal Light Awareness Program (FLAP), Ferro Corporation, Field Museum of Natural History, Florida Keys Audubon Society, Goldray Glass, Guardian Industries LLC, GZA GeoEnvironmental Inc., Hawk Mountain Sanctuary Association, Hong Kong University, Houston Audubon, Huge Company, Humane Society of the United States, Illinois Institute of Technology, International Crane Foundation, International Wildlife Rehabilitation Council, Italian League for Bird Protection (LIPU), Jacobsburg Environmental Education Center, Lacawac Sanctuary, Lafayette College, Lancaster County Bird Club, League of Women Voters Lehigh Valley, Lehigh Gap Nature Center, Lehigh Valley Audubon Society, Lehigh Valley Business, Lincoln Park Zoo, Linnaean Society of New York, Luther College, Manulife Corporation, Massey University, Millikin University, Ministry of the Environment and Climate Change (Canada), Monticello Bird Club, Mother Nature Network, Mountainaire Avian Rescue Society, Muhlenberg College, National Audubon Society, National Geographic Society, National Glass Association, National Institute of Ecology (Republic of Korea), National Public Radio, National Renewable Energy Laboratory, National Zoological Park (U.S.), Natural History Museum (Rotterdam), New York City Audubon, New York State Parks and Recreation, Niagara Falls State Park, Nippon Sheet Glass (Pilkington), Ontario Association of Architects, Orca Research Trust, Partners in Flight, Paul Smith College, Pennsylvania Game Commission, Pennsylvania Society of Ornithology, Philadelphia Zoo, Pittsburgh Plate Glass (Vitro), Pittsburgh

Acknowledgments

Plate Glass Foundation, Pleiotint Inc., Portland Bureau of Planning and Sustainability (Oregon), Public Broadcasting Service, Raptor Trust – Bird Rehabilitation and Education Center, Red Creek Wildlife Center, Rockridge Bird Club, Safari Club at Philadelphia Zoo, Safe Wings Ottawa, San Diego Zoo Global, San Francisco Planning Department, Saving Birds Thru Habitat – Charter Sanctuary, Science News, Smithsonian Migratory Bird Center, Solutia Inc., Southern Illinois University at Carbondale, Stockton University, Stony Brook University, Studio Gang, SurfaceCareUSA, Swarthmore College, Swift Parrot Recovery Team (Threatened Species Network, Australia), Swiss Ornithological Institute, Temple University, The Ecologist, The Nature Conservancy, Totman Exotic and Wildlife Rescue of Florida Keys Inc., Town of Brookhaven, Triton System Inc., University of British Columbia, University of Nebraska Kearney, University of Nebraska Lincoln, University of Technology Sydney, U.S. Fish and Wildlife Service, U.S. Green Building Council, View Dynamic Glass Inc., Villanova University, Virginia Tech University, Walker Glass Company, Washington and Lee University, Wildlands Conservancy, Wilkes University, Wilson Ornithological Society, Wisconsin Audubon Society, Wisconsin Humane Society, Whitehall Public Library, Wyncote Audubon Society, Zoological Lighting Institute.

WHO KNEW, NOT I, THAT you need an agent to be accepted by an established well-known publisher? The modest number but well-known successful agents I was referred to and asked to represent my book to publishers known to publish popular books in the biological science subjects told me they could not represent me because they judged there was no money in the topic that would interest those publishers. They did judge that in their opinion the subject was ideal for a university press. The few but well-known prestigious university presses I contacted informed me they too were not interested for the same reason; they informed me that such a book would not attract enough sales to equal their investment. Before and after signing a contract with Hancock House Publishers, I continued to rigorously seek a high image publisher based on recommendations of independent technical editors and my desire for what I felt would improve my ability to support this important avian conservation cause. I did this, remarkably, with Hancock's consent and encouragement. From my first interaction with them, they believed in the importance of the book, for the very values I wrote it: to help educate and thereby save countless bird lives from windows the world over.

I could not be more grateful to Hancock, especially the representative who consistently supported the book and its message, Myles Lamont. Thank you, Myles, for your support, understanding, and belief in this book as a worthy contribution to conservation science, education, and birds. Among others at Hancock working on producing this book who are worthy of my gratitude is editor Doreen Martens, designers Jana Rade, L. Raingam, and David Hancock.

TIMOTHY J. O'CONNELL AT OKLAHOMA State University was the formal scientific technical editor for this work. He deserves my highest praise and gratitude for his improvements and encouragement. Ellen Paul, former long-time, seemingly forever, Executive Director of the Ornithological Council volunteered to edit the manuscript. I eagerly accepted and she also deserves my highest praise and gratitude for her many improvements, especially those where her professional attorney knowledge and skills enhanced Chapter 10 dealing with the legal protection of birds. Both Ellen and Tim vastly improved the organization and language throughout the manuscript, in its seemingly infinite iterations. At various stages over literally decades, I received editorial improvements in content and language to what became this book from Daniel J., Heather Anne, Renee A., and Robyn L. Klem, Shawn Martell, and Peter G. Saenger. All the graphics were improved and design overall was by Robyn L. Klem. I thank them all for their invaluable contributions.

MY WIFE, RENEE A. (MUCCI) KLEM, and my former student, research assistant, and current ornithological science colleague Peter G. Saenger deserve special and extraordinary thanks from me and the countless bird lives they have saved during our work together studying birds, windows, and many other topics. They are among, if not exclusively, my best friends. Most of the logistical details of our bird-window investigations were handled effectively, efficiently, and expertly by Peter. Because of his enthusiastic attention to my modest contributions and professional reputation to this issue, given any appropriate opportunity, I regularly describe how Thomas H. Huxley was described as "Darwin's Bulldog." With as much energy, if not far more, Peter G. Saenger qualifies as "Klem's Bulldog." None of what I have been able to accomplish, personally or professionally, would have been possible without Renee. Although we have been separated many times because of my work, we are inseparable. She prepared almost all the materials for my very first experiments, and continues to the present to offer any help whenever needed. Her encouragement, all manner of personal and professional support, to include direct intervention, was required to get me through and complete my doctoral degree focusing on the subject of this book. Please permit and tolerate my repeating for emphasis, none of what I have been able to fulfill on behalf of the birds we both care so much about would be possible without her. Thank you for all you have done, all that you have and continue to mean to me, and for being the uniquely special human being you are Renee.

IT MAY TAKE A VILLAGE to raise a child, but it took legions of helpers to assist me in my work on behalf of birds. This book could not have been written without each and every one of them, no matter how small or large the contribution, but all errors and omissions herein are mine alone. For all those who have read this far, I deeply apologize if I failed to list your name as a small token of my thank you for your help.

Index

architects   13, 26, 62, 98-101, 116, 127, 144, 147, 154
Biological Survey   136

Colleges and Universities
Alfred University   119
Augustana College   57, 84
Aurora University   84
College of William and Mary   107
Earlham College   64
Hofstra University   14, 20, 47
Loyola College   95
McGill University   15, 104
Memorial University   104
Millikan University   84
Moravian College   96
Muhlenberg College   14, 46, 89, 94-95, 117, 124, 131, 146
Northeastern University   19
Northwestern University   97
Pennsylvania State University   84
Principia College   84
Southern Illinois University at Carbondale   14, 20, 22, 43
Swarthmore College   95, 97, 117
Swedish University of Agricultural Sciences   117
Tufts University   69
University of Alberta   66
University of Birmingham   117
University of British Columbia   145
University of Illinois   100
University of Kansas   59
University of Nebraska State Museum   76
Uppsala University   118
Valley City State College   30
Wesleyan University   83
Wilkes University   14, 18

Collision Listserve   101

Companies
3M   105, 115-116
Acopian BirdSavers   114-115, 144, 157
AGC of North America   105
Air Products and Chemicals   128
Arnold Glas   105, 124, 126-127
Brigham Oil and Gas   138
Cadillac Fairview Corporation   141
CollidEscape   105, 115-116, 158
Convenience Group   105, 116
CPFilms   97, 121-124, 127-130
Design IP   128
Eastman Chemical Company   129
Ecojustice   141
Erickson International   105
Feather Friendly   116, 157
Fox and Fowle   124
Goldray Glass   105
Nippon Sheet Glass (NSG)   105
Pilkington   105

Pittsburgh Plate Glass (PPG)   49-51, 105
Studio Gang Architects   100
Surfacecareusa   105, 122
Vitro   105
Walker Glass   99, 105, 131, 133

Environmental Threats

attractants   62-63, 107-108, 146
cats   12, 34-36, 39
Communication towers   35, 37, 101-102
DDT   40, 79, 141
Exxon Valdez   34, 148
Gulf oil spill   34
noise barriers   26, 63, 100, 114, 132
oil spills   34, 148
pesticides   35, 79
wind turbines   36, 102, 137, 140

Injuries

brain tissue   68
concussions   71
femur   70
furcula   70
head trauma   9, 12, 24, 68, 70
histological sections   70
humerus   70
injuries   12, 25, 47, 68-72, 74, 81-83, 152-154
sternum   70
tibia   70
ulna   70

Kimya   146

Legislation, Regulation

Bird Directive   134
Bird Friendly Design Ordinance   142
Endangered Species Act (ESA)   134, 140
Environmental Protection Act (EPA)   134
House Bill (H.B.) 2   141
H.B. 919   141
H.B. 1643   141
H.B. 2078   141
H.B. 2280   141
H.B. 2542   141
H.B. 4797   141
Important Bird Area (IBA)   93
Leadership in Energy and Environmental Design (LEED)   101, 145
North American Landbird Conservation Plan   36
Species at Risk Act (SARA)   134
Weeks-McLean   136
U.S. Patent   123, 125, 128

Mitigation Methods

2 x 2 Rule   133, 149-150, 157
2 x 4 Rule   112, 114, 116, 133, 149-150
acid etch   116, 133, 149
Acopian BirdSavers   114-115, 144, 157
AviProtek T   131, 133
ceramic frit   116-117, 133, 149
CollidEscape   105, 115-116, 158
CPFilms   97, 121-124, 127-130
Feather Friendly   116, 157

feeder placement   13, 108
field experiments   49, 52-55, 110-111, 113-114, 124, 126
flight cage   42, 49, 52-56
nanoparticles   119
Ornilux Mikado   124-128
Surface #1   63, 111, 114-115, 124, 131-132, 149
tunnel testing   42, 52, 110, 113-114, 124-126

News Media Outlets

Associated Press (AP)   95
British Broadcasting Corporation (BBC)   89, 92
Chicago Sun-Times   88
CNN   95
Migratory Bird Conservation Act (MBCA)   134
Migratory Bird Treaty Act (MBTA)   102, 134-135
Minnesota Vikings   62, 138
Missouri v. Hollander   136
Montreal Gazette   105
National Public Radio (NPR)   97, 121

North American Ornithological Conference (NAOC)   85

Organizations, Agencies

Non-governmental

Acopian Center for Ornithology   46
American Bird Conservancy (ABC)   100-101, 115, 125-126, 148
American Institute of Architects (AIA)   99-100, 138
American Museum of Natural History (AMNH)   48, 73, 89
American Ornithologists' Union (AOU)   73, 89, 104
Association of Field Ornithologists (AFO)   86
Audubon Pennsylvania   146
Avian Care and Research Foundation   69
Bird Safe Glass Working Group   102, 104
Bird-window Collision Working Group (BCWG)   24, 146, 150
British Ornithologists' Union (BOU)   104
Canadian Society of Ornithology (CSO)   104
Canadian Standards Association (CSA)   143
Center for Biological Diversity   139
Chicago Academy of Sciences   88
Chicago Bird Collision Monitors   98, 150-151
Chicago Field Museum   76, 100
Chicago Ornithological Society   97, 151
Cleveland Museum of Natural History   74
Cornell Lab of Ornithology   90, 93
Dead Duck Society   44
Defenders of Wildlife   139
Detroit Zoological Society (DZS)   145-146

Evanston North Shore Bird Club   97
Fatal Light Awareness Program (FLAP)   60-61, 92, 98, 101, 127, 148, 150-151, 155
Hawk Mountain Sanctuary Association   90-91, 115-116
Illinois Institute of Technology (IIT)   98, 100
Lehigh Valley Audubon Society   146, 159
Max Planck Institute for Ornithology   124
Migratory Bird Center   34, 84
Moon Lake Electric Association   138
National Audubon Society (NAS)   40, 90, 99, 130, 139, 146
National Geographic   49, 89
New York City Audubon   84, 101-102, 104, 106, 150-151
Portland Audubon   150
Raptor Research   48
Raptor Trust   69
Smithsonian Conservation Biology Institute   34, 85
United States Green Building Council (USGBC)   101, 144
Virginia Society of Ornithology   84
Waterbird Society   92
Wild Bird Feeding Industry (WBFI)   84
Wild Birds Unlimited   101
Wildlife Cooperative Research Laboratory   52
Wilson Ornithological Society   73, 82-83, 86
World Wildlife Fund Canada (WWF)   100
Wyncotte Audubon Society   146

Governmental

Association for Environment and Nature Conservation Germany   142
Canadian Wildlife Service (CWS)   34-35
Chicago Department of the Environment   98
Department of the Interior (DOI)   139
Division of Natural Heritage (DNH)   84
Environment Canada   34-35, 84
Internal Revenue Service (IRS)   50
National Zoological Park   85
New York State Parks   92
United States Fish and Wildlife Service (USFWS)   79, 102-104, 137-140

Ornithology   21

People

Aborn, David   73
Acopian, Jeff   115
Acopian, Sarkis   96
Allen, Deborah   60
Altadonna, Leigh   146
Anthony, Terence R.   43
Arnold, Hans-Joachim   124
Arnold, Todd   37

Index

Attenbourgh, Sir David 92
Babcock, Judge 138
Baird, Spencer F. 24
Baker, Andy 145
Banks, Richard C. 43
Barlow, Jon C. 43
Bartell, Joe 106
Bauer, Mrs. 67
Bayne, Erin 66, 85
Beason, Robert 31
Beecher, William J. 88
Bent, Arthur Cleveland 59
Berthold, Peter 67
Bird, David M. 104
Blahous, C. P. 50
Blancher, Peter 34
Boano, Giovanni 85
Bonney, Rick 90-91
Boswall, Jeffery 89
Bracey, Annie 66
Bradley, Senator Bill 131-132
Brereton, Raymond 40
Brewer, Thomas M. 24
Brewster, William 24
Brown, G. M. 45
Buckley, Paul A. 20
Burtt Jr., Edward (Jed) H. 83
Butler, Michael 73
Campenhausen, Mark von 31
Carpenter, F. 65
Carson, Rachel 79
Choudhory, Sharmila 92
Chubb, Kit 69-70
Chubb, Robin 69
Clark, Joe 91
Clench, Mary H. 43
Coolidge, Calvin 81-82

Corcoran, Larry Martin 135
Cortes, Rui 85
Cotton, Karen Imparato 101-102, 151
Craig, Alicia Francis 101
Craig, Matthew 58
Creshkoff, Rebekah 98, 106, 151
Cusa, Marine 57, 62
Dangerfield, Rodney 93
Dann, Donnie 97
Darwin, Charles 83
DeCandido, Robert 60
Delacretaz, Nicole 62, 84
Dickerson, Mr. and Mrs. Dwayne 47
Dinetti, Marco 85
Doeker, Randi 97-99, 112, 151
Dunn, Erica 58
Emlen Jr., John 30
Farrel, W. M. 30
Fernandes, Luis Sanches 85
Finkelstein, Erica 94
Fraser, David 145
Gang, Jeanne 100
Garces, Andreia 85
Gelb, Yigal 62, 84
Gelman, Ben 88
George, William G. 14, 20-21
Gibson, Eleanor 30
Gill, Frank B. 37
Goldsmith, Timothy 31
Graber, Jean and Dick 64
Green, Judge Melvyn 141
Grimes, Ellen Dineen 100
Guo-Qiang, Cai 104
Hager, Stephen B. 57-58, 62, 84
Hall, G. A. 65

Harwood, Michael   90
Hastad, Olle   31, 117-118
Haupt, Heiko   85
Hayward, Jack and Muriel   44-45, 47
Hill, Libby   97
Hodos, William   31
Horn, David J.   84
Horton, Frank E.   51
Hovland, Judge   138-139
Humboldt, Alexander von   77
Hunsinger, Robbie Lynn   98, 151
Ingrassia, Nicole Marie   107
Johns, Jamie   152
Johnson, Lyndon B.   19
Johnston, Richard   59
Jones, Andy   74
Joseph, Joan E.   145
Kagan, Ron   145
Keating, Kiki   94-95
Kenney, Devin T.   135
King, Alicia Francis   101
Kirschfeld, Kuno   31
Klem, Renee   92, 124
Klimstra, William D.   20, 48, 52
Koehl, Albert   141
Kolbert, Elizabeth   83
Konig, Claus   30, 65
Lambertucci, Sergio A.   85
Langridge, H. P.   29
Lanyon, Wesley   48
Larson, Gary   86
Leffel, Robert   109
Lewis, William M.   51
Lincoln, Abraham   123
Liviglio, Joann   95
Lohrl, Hans   59

Loio, Sara   85
Longcore, Travis   77
Loss, Scott R.   34-35, 37, 39-40, 62, 84, 144, 151
Lovell, H. B.   65
Lyell, Charles   83
Lyons, Thomas   92
MacGregor-Fors, Ian   85
Machtans, Craig S.   35, 84
Malakoff, David   96
Manning, Jeffery   122
Manville II, Albert M.   102
Martin, Graham   31, 117
Martin, Joe   86, 88
McAdams, E. J.   102
Meltz, Amanda (Mandy)   145-146
Menacho-Odio, Rose Marie   85
Mesure, Michael   150-151
Meyerhuber, Alfred   124-125
Moenich, Chris   88
Montiani, Fabiano   85
Nielson, John   97
Nuttall, Thomas   23
Nyffleer, Martin   78
Ocompo Penuela, Natalia   85
O'Connell, Timothy J.   39, 84
Odeen, Anders   31, 117-118
Olson, Wendy   98
Pacheco, Fernando   85
Parker, Carolyn   98
Parkins, Kaitlyn   65
Pettingill Jr., Olin Sewall   93
Piratelli, Augusto   85
Pires, Isabel   85
Pokras, Mark   69
Pollock, Judy   100

Port, Anthony (Tony) Brian 97, 121, 129
Post, S. 64
Potter, Rene 48, 65
Prada, Justina 85
Price, Steven 100-101
Prince, Annette 98, 151
Queiroga, Felisbina 85
Quigley, Mike 141
Rebolo-Ifran, Natalia 85
Ridgeway, Robert 24
Robbins, Chandler 36
Rogers, Steven 29
Root, Senator Elihu 136
Rosen, Efrem 47
Rosenberg, Kenneth V. 37
Rossler, Martin 85, 113
Rothstein, Poly 47
Russell, Keith 146
Sabo, Ann 59
Saenger, Peter G. 26-27, 39, 46, 55, 111, 125, 146
Sayre, Roxanna 90
Schendel, Mark 100
Schmoyer, Irvin R. 120
Sheppard, Christine 101, 110, 113, 151
Simmons, James C. 128-129
Sinner, Paul 30
Skutch, Alexander F. 79
Smith, P. A. 77
Soeiro, Vanessa 85
Soucy, Len 69
Stotz, Douglas F. 100
Tallarico, R. B. 30
Talpin, Corella A. 76
Thompson, William H. 91
Thurber, James 9
Townsend, Charles W. 24, 45
Trudell, Heidi 84
Valum, Brynjulf 64
Van Dam, Bonnie 145
Veltri, Carl 69
Virgilio, Agustina di 85
Voegler, Grace K. 50-51
Walk, Richard 30
Wall, Gordon Lynn 28
Walley 29
Walls, George 31
Watterson, Bill 86-87
Willard, David 100
Willet, George 28
Williams, John I. 94
Williams, Timothy 97
Witzler, S. 64
Yakutchik, Maryalice 95
Yambert, Paul and Carla 44
Zbyryt, Adam 85
Zink, Robert 37

Places

Countries, Cities and States

Austria 85, 113-114, 160
Barrington, Illinois 143
Beltsville, Maryland 86
Bronx Zoo 101, 124
Calgary, Alberta 143
California 28, 129, 140, 142-143
Canada 12, 25, 34-35, 43, 61, 85, 99-100, 102, 105, 116, 131, 134-135, 148, 150, 155
Carbondale, Illinois 14, 43-45, 58
Chicago, Illinois 60, 76, 88, 97-100, 102-103, 141, 150-151

Denver, Colorado   106
Detroit, Michigan   106
Duluth, Minnesota   66
San Salvador, El Salvador   76
European Union   134
Evanston, Illinois   97, 99
Germany   25, 65, 85, 105, 125, 142, 160
Illinois   30, 44-48, 50, 52, 56-58, 60, 62-66, 72-74, 84, 140, 142-143
Jamaica Plains, New York   24
Kentucky   65
Makanda, Illinois   48, 65
Martinsville, Virginia   97, 121-122
Minneapolis, Minnesota   62, 115
Netherlands   25, 44, 161
Newfoundland   104
New York City   60, 62, 65, 98, 124
New Zealand   161
Niagara Falls, New York   92-95
North Dakota   30, 138
Palo Alto, California   129, 143
Philadelphia, Pennsylvania   95, 145-146, 150
Portland, Oregon   142-143, 150
Rock Island, Illinois   57, 84
St. John's, Newfoundland   104
Tennessee   73
Toronto, Ontario   57, 61-62, 65, 92-93, 98, 106, 141-143, 150
Virginia   59, 65, 84, 97, 109, 121-122
Washington D. C.   98, 101, 150

Buildings

Empire State Building   60
Ford Calumet Environmental Center   100
Independence National Historic Park   145-146
McCormick Place   88, 99-100, 103
Metropolitan Museum of Art   104
Neckers chemistry building   21-22
New York Times building   109
Philadelphia Zoo   145
Soldier Field   99
Toronto Zoo   116
U.S. Bank Stadium   62
World Trade Center   106

Publications

A Bird Watcher's Adventures in Tropical America   79
A History of North American Birds   24
A humorous look at a deadly conservation issue: birds and glass   86
A Manual of the Ornithology of the United States and Canada   23
American Birds   65, 76
Annual Review of Ecology, Evolution and Systematics   37
Audubon magazine   90, 96
Auk   24, 43, 45, 82
Avian Conservation Ecology   77
Behaviour   30
Biologia   113
Biological Conservation   151

Index

Bird Watcher's Digest   91, 105
Clear & Present Danger   96
Collision Course: The Hazards
    of Lighted Structures and
    Windows to Migrating
    Birds   101
Colorado Field Ornithology (C.F.O.)
    Journal   29
Condor   28, 35, 40, 82-83, 144
Condor: Ornithological
    Applications   35, 40, 144
Conservation Biology   40
Corella   76
Denver University Law
    Review   135
Deutsche Sektion des
    Internationalem Rates für
    Vogelschultz Bericht   30, 65
Fatal reflections   95
Florida Naturalist   29
Glass: a deadly conservation issue
    for birds   85
Ibis   117
Journal of Animal and Natural
    Resource Law   135
Journal of Comparative
    Physiological Psychology   30
Journal of Field Ornithology   58,
    68, 83
Journal of the Pennsylvania
    Academy of Science   76
Kentucky Warbler   65
Kosmos   59
Living Bird Quarterly   91
Nature Communications   34
North American Birds   24, 76
Northeastern Naturalist   62, 65

PeerJ   58-59, 118
Ornithology   23, 37, 39
Philadelphia Inquirer   95
PLoS One   37, 62
Popular Science   91
Psychological Monograph   30
Raven   84
Red List   40
Reducing Bird Collisions with
    Buildings and Building Glass
    Best Practices   104
Science   37, 82
Science of Nature   78
Scientific American   30
Silent Spring   79
Sixth Extinction   83
Southern Illinoisan   88
State of the Birds   36
Sterna   64
The Avian Ark: Tales from a Wild-
    bird Hospital   70
The Kingbird   60
Urban Ecosystems   57, 62
Vertebrate Eye and Its Adaptive
    Radiation   28
Watchlist   40
Wildlife Research   66
Wilson Bulletin   82, 109
Wilson Journal of
    Ornithology   57, 62, 66, 82,
    122, 125, 128
Pulitzer Prize   83

Species

Blackbird, Eurasian   44
    Red-winged   85
Blackcap   44, 67

Bobwhite, Northern   40
Bristlebird, Eastern   40
Bullfinch, Eurasian   44
Bunting, Indigo   73
  Painted   40
Cardinal, Northern   26, 44, 48, 63
Chaffinch   44
Chat, Yellow-breasted   71-72
Chickadee, Black-capped   67
Cockatoo, Yellow-crested   40
Creeper, Brown   76
Cricket, Mormon   77
Cuckoo, Yellow-billed   24, 45, 63
Darter, Oriental   40
Dove, Mourning   21, 70
Emerald, White-bellied   76
Falcon, Peregrine   69
Finch, House   56
Firetail, Diamond   40
Flicker, Northern   44
Flycatcher, Olive-sided   40
  Pied   44
Gannet, Cape   40
Goldcrest   44
Goose, Canada   96
Goshawk, Northern   70
Grassbird, Marsh   40
Greenfinch, European   44
Green-pigeon, Whistling   40
Grosbeak, Evening   44, 73
  Rose-breasted   75
Grouse, Ruffed   70
Gull, California   77
Hawk, Cooper's   70
  Sharp-shinned   27, 70
Hummingbird, Ruby-throated   40, 44, 63, 97, 146

Jay, Blue   57
Junco, Dark-eyed   44, 48, 51, 53, 63
Kestrel, American   56-57
Kinglet, Golden-crowned   57
Mallard   44, 119
Nuthatch, White-breasted   41
Ovenbird   40, 44, 57, 69
Parrot, Superb   40
  Swift   40
Petrel, Gould's   40
Pheasant, Copper   40
Pigeon, Plain   40
  Rock   58
Rail, Black   40
  Virginia   72
Robin, American   38, 44, 51, 58, 70-71, 76
  European   44
  Flame   40
Sapsucker, Yellow-bellied   40, 44, 85
Shearwater, Townsend's   40
Sparrow, Brewer's   40
  House   58, 76
  Lincoln's   76
  Song   56
  White-crowned   63, 69
  White-throated   44, 48, 57, 71-72
Sparrowhawk, Eurasian   44
Starling, European   58
Storm-Petrel, Band-rumped   44
Swallow, Barn   59
Thick-knee, Bush   40
Thrush, Hermit   44
  Song   44

Swainson's   28-29, 44
Wood   40-41
Tit, Great   44
Vireo, Bell's   40
   Blue-headed   57
   Red-eyed   58
   White-eyed   63
Warbler, Bay-breasted   51
   Blackpoll   78-79
   Cerulean   40
   Garden   44
   Golden-winged   40
   Hooded   73
   Kentucky   40
   Kirtland's   40
   Nashville   63, 75
   Tennessee   44
   Willow   44
   Worm-eating   40
   Yellow-rumped   44
Waxwing, Cedar   29, 44, 59, 63

Woodcock, Eurasian   44
Woodpecker, Downy   70
   Hairy   20, 70
   Red-bellied   48
   Red-headed   40

Vision

   accommodation   31
   aposematic coloration   120
   cone cells   31-32
   cornea   31
   fovea   31
   lens   31-32
   oil droplets   32
   retina   31-32
   rod cells   31
   spectrophotometer   121, 127
   tetrachromatic   32
   ultraviolet (UV)   97, 117-133
   visual spectrum   120

window-washers   106

About the Author

DANIEL KLEM, JR. IS SARKIS Acopian Professor of Ornithology and Conservation Biology in the Department of Biology at Muhlenberg College, Allentown, Pennsylvania. He and his wife Renee A. (Mucci) Klem are the proud parents of Heather Anne, Robyn Lynne, and Daniel Joseph Klem. He grew up in Larksville, Edwardsville, and Kingston within the Wyoming Valley and surrounding mountains of northeastern Pennsylvania. He was formally educated in the Larksville School District, Saint Hedwigs Roman Catholic Church School (Edwardsville), Kingston School District (1964, Kingston High School), Wilkes College (1968, B.A. Biology, now Wilkes University), Wilkes-Barre, Pennsylvania, Hofstra University (1973, M.A. Biology, Hempstead, New York), and Southern Illinois University at Carbondale (1979, Ph.D.). He performed distinguished service in the United States Army from 1968-1971. For the last 47 years and continuing to the present he studies, writes, and teaches about birds. His most prestigious awards are the students and professional colleagues he has interested in bird study. He guided and encouraged Sarkis Acopian of Easton, Pennsylvania to financially support the establishment of the Acopian Center for Ornithology at Muhlenberg College, and the Acopian Center for Conservation Learning at the Hawk Mountain Sanctuary Association, Kempton, Pennsylvania. He led and co-authored the scientific part of the Birds of Armenia Project (BOA) whose principal goal was to use bird study to promote an environmental ethic and the preservation of natural lands in the Republic of Armenia from 1992-2000. Additionally, BOA brought previously unavailable scientific ornithology from Armenia and the greater Caucus region to the western scientific community. The BOA Project published *A Field Guide to the Birds of Armenia* (1997), an abbreviated Armenian language field guide version (2000), and a *Republic of Armenia Birds of Armenia Reference Map* (1999) for the general reader, and the scholarly *Handbook of the Birds of Armenia* (1999) for scientific ornithologists and naturalists. Most of his current research consists of using detailed observations and experiments to evaluate novel bird-window collision deterrent prototypes, preparing review papers describing the historic and lastest means to save birds from windows, and serving as a technical consultant to glass manufacturers developing bird-safe sheet glass, various national and international conservation organizations, and local, regional, and federal government agencies the world over.

related titles by Hancock House Publishers

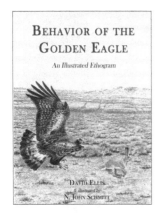

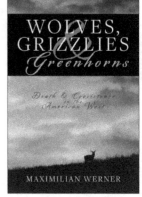

Behavior of the Golden Eagle: *an illustrated ethogram*
David Ellis, 2017
ISBN 978-0-88839-078-3 Trade SC
102pp, 88 illustrations, 8½ x 11

North Amer. Ducks, Geese & Swans: *identification guide*
Frank Todd, 2018
ISBN 978-088839-093-6 Trade SC
208pp, 1000+ photos, 6½ x 9½

Wolves, Grizzlies & Greenhorns: *death & co-existence in the American West*
Maximillian Werner, 2021
ISBN 978-0-88839-537-5 Trade SC
352pp, 7 photos, 5½ x 8½

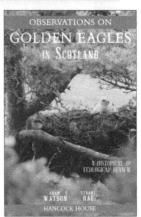
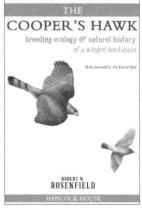

Observations of Golden Eagles in Scotland
Adam Watson & Stuart Rae, 2019
ISBN 978-0-88839-030-1 Trade SC 146pp, 116 photos, 6 x 9

The Cooper's Hawk: *breeding ecology & natural history of a winged huntsman*
Robert Rosenfield, 2018
ISBN 978-0-88839-082-0
Trade SC
164pp, 73+ photos, 6 x 9

Raptor Research & Management Techniques
David Bird & Keith Bildstein, 2007
ISBN 978-0-88839-639-6 SC
464pp, 66 photos, 8½ x 11

Hancock House Publishers
19313 Zero Ave, Surrey, BC V3Z 9R9
www.hancockhouse.com
info@hancockhouse.com
1-800-938-1114